T0210902

Lecture Notes in Computer Science 11290

Commenced Publication in 1973
Founding and Former Series Editors:
Gerhard Goos, Juris Hartmanis, and Jan van Leeuwen

More information about this series at http://www.springer.com/series/8851

Ngoc Thanh Nguyen · Richard Kowalczyk
Jacek Mercik · Anna Motylska-Kuźma (Eds.)

Transactions on Computational Collective Intelligence XXXI

 Springer

Editor-in-Chief
Ngoc Thanh Nguyen
Department of Information Systems
Wrocław University of Technology
Wrocław, Poland

Co-Editor-in-Chief
Richard Kowalczyk
Swinburne University of Technology
Hawthorn, VIC, Australia

Guest-Editors
Jacek Mercik
WSB University Wroclaw
Wroclaw, Poland

Anna Motylska-Kuźma
WSB University Wroclaw
Wroclaw, Poland

ISSN 0302-9743   ISSN 1611-3349 (electronic)
Lecture Notes in Computer Science
ISSN 2190-9288   ISSN 2511-6053 (electronic)
Transactions on Computational Collective Intelligence
ISBN 978-3-662-58463-7   ISBN 978-3-662-58464-4 (eBook)
https://doi.org/10.1007/978-3-662-58464-4

Library of Congress Control Number: 2018958639

This Springer imprint is published by the registered company Springer-Verlag GmbH, DE part of Springer Nature
The registered company address is: Heidelberger Platz 3, 14197 Berlin, Germany

Transactions on Computational Collective Intelligence
Vol. 31

Preface

It is my pleasure to present to you the 31st volume of LNCS *Transactions on Computational Collective Intelligence*. In Autumn 2017 (November 24) at the WSB University in Wroclaw, Poland, the third seminar on "Quantitative Methods of Group Decision-Making" took place. Thanks to WSB University in Wroclaw we had an excellent opportunity to organize and financially support the seminar. This volume presents the post-seminar papers by participants at this seminar. During the seminar, we listened to and discussed over 19 presentations from 16 universities. The present volume contains 12 high-quality, carefully reviewed papers.

The first paper "An Equivalent Formulation for the Shapley Value" by Josep Freixas is devoted[1] to the study of an equivalent explicit formula for the Shapley value. Its equivalence with the classic one is proven by double induction. The importance of this new formula, in contrast to the classic one, is its capability of being extended to more general classes of games, in particular to j-cooperative games or multichoice games, in which players choose among different levels of participation in the game.

In the second paper entitled "Reflections on Two Old Condorcet Extensions" by Hannu Nurmi, the authors report that the distinction (in fact, rivalry) between two intuitive notions about what constitutes the winning candidates or policy alternatives has been present in the social choice literature from its Golden Age, i.e., in the late eighteenth century. According to one of them, the winners can be distinguished by looking at the performance of candidates in one-on-one, that is, pairwise contests. According to the other, the winners are in general best situated in the evaluators' rankings over all candidates. The best known class of rules among those conforming to the first intuitive notion are those that always elect the Condorcet winner whenever one exists. These rules are called Condorcet extensions for the obvious reason that they extend Condorcet's well-known winner criterion beyond the domain where it can be directly applied. A candidate is a Condorcet winner whenever it defeats all other candidates in pairwise contests with a majority of votes. Condorcet extensions specify winners in all settings including those where a Condorcet winner is not to be found. Of course, in those settings where there is a Condorcet winner they all end up electing it. The focus of this paper is on Condorcet extensions, more specifically on two choice

---

[1] The description of the papers are taken from summaries prepared by their authors.

rules dating back to the nineteenth century, namely, Dodgson's and Nanson's procedures.

In the third paper "Transforming Games with Affinities from Characteristic into Normal Form," by Cesarino Bertini, Cristina Bonzi, Gianfranco Gambarelli, Nicola Gnocchi, Ignazio Panades, and Izabella Stach present a study of a method for transforming games in extensive form into normal form. John von Neumann and Oskar Morgenstern, while introducing games in extensive form, also supplied a method for transforming such games into normal form. Once more, the same authors provided a method for transforming games from characteristic function form into normal form, although limited to constant-sum games. Gambarelli proposed a generalization of this method to variable-sum games. In this generalization, the strategies are the requests made by players to join any coalition, with each player making the same request to all coalitions. Each player's payment consists of the player's request multiplied by the probability that the player is part of a coalition truly formed. Gambarelli introduced a solution for games in characteristic function form, made up of the set of Pareto-Optimal payoffs generated by Nash equilibria of the transformed game. In this paper, the transformation method is generalized to the case in which each player's requests vary according to the coalition being addressed. Propositions regarding the existence of a solution are proved. Software for the automatic generation of the solution is supplied.

The fourth paper "Comparing Game-Theoretic and Maximum Likelihood Approaches for Network Partitioning" by Vladimir Mazalov shows the relationship between the game-theoretic approach and the maximum likelihood method in the problem of community detection in networks. The formulation of a cooperative game related to a network structure where the nodes are players in a hedonic game is given and then the stable partition of the network into coalitions is found. This approach corresponds to the problem of maximizing a potential function and allows one to detect clusters with various resolution. The maximum likelihood method for the tuning of the resolution parameter in the hedonic game is proposed and illustrated by some numerical examples.

In the fifth paper entitled "Comparing Results of Voting by Statistical Rank Tests," Honorata Sosnowska and Krzysztof Przybyszewski consider and analyze the results of voting by different voting methods. Approval voting, disapproval voting, categorization method, and classic majority voting are compared using results of a presidential poll conducted over a representative sample. Results differ depending on the voting methods used. However, from the statistical point of view, within the same valuation scope (positive if voting is constructed so that the vote is "for" the candidate and negative if the vote is "against") the order of the candidates remains similar across the methods.

In the sixth paper "Remarks on Unrounded Degressively Proportional Allocation," Katarzyna Cegiełka, Piotr Dniestrzanski, Janusz Łyko, and Arkadiusz Maciuk analyze the degressiveness as a principle of distributing indivisible goods. The Lisbon Treaty has legally endorsed degressiveness as a principle of distributing indivisible goods. Yet the principle has been implemented in executive acts with insufficient precision. As a result, it cannot be unambiguously applied in practice. Therefore many theoretical studies have been conducted aiming at a more precisely defined formulation of the principle so that resulting allocations can be explicitly derived from primary rules. This paper aims at submitting a formal definition of unrounded degressively proportional distribution.

In the seventh paper entitled "On Measurement of Control in Corporate Structures," Jacek Mercik and Izabella Stach investigate various methods to measure power control of firms in corporate shareholding structures. This paper is a study of some game-theoretical approaches for measuring direct and indirect control in complex corporate networks and it is concentrated on the comparison of methods that use power indices to evaluate the control-power of firms involved in complex corporate networks. They analyze the Karos and Peters as well as the Mercik and Lobos approaches. In particular, they assess the rankings of firms given by the considered methods and the meaning of the values assigned by the power indices to the stock companies presented in corporate networks. Some new results were obtained. Specifically, taking into account a theoretical example of a corporate shareholding structure, the authors observe the different rankings of investors and stock companies. Then, some brief considerations about the reasonable requirements for indirect control measurement are provided, and ideas of modifying the implicit index are undertaken. The paper also provides a short review of the literature of the game theoretical approaches to measure control power in corporate networks.

The eighth paper entitled "The Effect of Brexit on the Balance of Power in the European Union Council Revisited: A Fuzzy Multicriteria Attempt" by Barbara Gładysz, Jacek Mercik, and David Ramsey presents the approaching exit of Great Britain from the European Union, which raises many questions about the changing relations between other member states. In this work, the authors propose a new fuzzy game for multicriteria voting. They use this game to show changes in Shapley's values in a situation where the weights of individual member countries are not determined and they describe non-determinism with fuzzy sets. In particular, this concerns considerations related to pre-coalitions.

In the ninth paper "Robustness of the Government and the Parliament, and Legislative Procedures in Europe" by Chiara De Micheli and Vito Fragnelli, the analysis of the procedures of the Italian Constitution, focussing on their strength and correlating it with the strength of the government and of the parliament, measured through two parameters, governability and fragmentation, is presented. The authors also extend the analysis to other European democracies: the UK, France, and Spain.

The tenth paper entitled "Should the Financial Decisions Be Made as Group Decisions?" is written by Anna Motylska-Kuzma. The aim of the article is to verify the effectiveness of financial decision-making as group decisions. The research consisted of a simple simulation game conducted with a group of students. There were 540 single games analyzed, taking into account the decisions and their effectiveness with regard to the size of the decision group. ANOVA, dependent paired samples test, as well as independent samples test were used to verify the hypothesis. The studies show that simple decisions about production and price are almost the same, regardless of size of the group, but the capital structure differs significantly between the groups and between the individuals and the group. The results also indicate differences in the effectiveness of the decision-making. The findings can be used to enhance the performance of the decision-making process in companies, especially within the scope of finance.

The 11th paper is the joint work of David Ramsey, Anna Kowalska-Pyzalska, and Karolina Bienias and is entitled "Diffusion of Electric Vehicles: An Agent-Based Modelling Approach." In recent years, most European governments have clearly stated their aims to promote the production and sale of electric vehicles (EVs), which are seen to be an environmentally friendly means of transport. At the beginning of 2018, the Polish government introduced legislation to promote the diffusion of electric vehicles. In conjunction with this, the authors present some preliminary work on a major study on the diffusion of EVs on the Polish market. This project aims to simulate the behavior of the market using an agent-based model, which will be based on the results of a survey carried out among the purchasers of new cars in two major Polish cities, Wroclaw and Katowice. Agent-based models allow the decision of agents to be affected by interactions with neighbors. The diffusion of EVs in Poland is at its very beginning, but according to experts the perspectives for further market development are promising. This article describes some of the basic factors behind a household's decision to buy an EV. A simple model is presented, together with a discussion of how such a model will be adapted to take into account the results from the study.

The 12th paper "Decision Process Based on IOT for Sustainable Smart Cities" is written by Cezary Orłowski, Arkadiusz Sarzyński, Kostas Karatzas, and Nikos Katsifarakis. The work presents the use of decision trees for the improvement of building business models for IoT (Internet of Things). During the construction of the method, the importance of decision trees and business models for decisions-making was presented. The method of using SaaS (Software as a Service) technology and IoT is proposed. The method has been verified by applying it to building business models for the use of IoT nodes to measure air quality for smart cities.

I would like to thank all authors for their valuable contributions to this issue and all reviewers for their opinions, which helped to ensure the high quality of the papers. My very special thanks go to Prof. Ngoc-Thanh Nguyen, who encouraged us to prepare this volume and who helped us publish in due time and in good order.

August 2017   Jacek Mercik
                                                        Anna Motylska-Kuźma

Transactions on Computational Collective Intelligence

This Springer journal focuses on research in applications of the computer-based methods of Computational Collective Intelligence (CCI) and their applications in a wide range of fields such as semantic web, social networks and multi-agent systems. It aims to provide a forum for the presentation of scientific research and technological achievements accomplished by the international community.

The topics addressed by this journal include all solutions of real-life problems, for which it is necessary to use computational collective intelligence technologies to achieve effective results. The emphasis of the papers published is on novel and original research and technological advancements. Special features on specific topics are welcome.

Contents

An Equivalent Formulation
for the Shapley Value

Josep Freixas[✉]

Universitat Politècnica de Catalunya
(Campus Manresa i Departament de Matemàtiques),
08242 Manresa, Spain
josep.freixas@upc.edu

Abstract. An equivalent explicit formula for the Shapley value is provided, its equivalence with the classical one is proven by double induction. The importance of this new formula, in contrast to the classical one,
is its capability of being extended to more general classes of games, in
particular to j-cooperative games or multichoice games, in which players
choose among different levels of participation in the game.

Keywords: Cooperative games · Marginal contributions
Shapley value · Alternative formulations · Potential generalizations

1 Introduction

The Shapley value, see Shapley (1953), (1962), admits a clear formulation in
terms of marginal contributions. As Shapley described, his value is based on the
following model: (1) starting with a single member, the coalition adds one player
at a time until everyone has been admitted; (2) the order in which the players
are to join is determined by chance, with all arrangements equally probable;
(3) each player, on her admission, demands and is promised the amount which
her adherence contributes to the value of the coalition as determined by the
characteristic function, i.e., the marginal contribution. Such model was criticized
by several authors as highly artificial (see e.g. Luce and Raiffa (1957) and Brams
(1975) among others).

Instead, step (3) could be replaced by the following ones:

(3.1) In her turn, each player decides whether to cooperate or not in forming a
      proposed coalition.
(3.2) If in her turn the player has decided to cooperate, then she receives the
      marginal contribution of the coalition formed by those players preceding
      her that decided to cooperate.
(3.3) If in her turn the player has decided not to cooperate, then she receives the
      marginal contribution of the coalition formed by those players preceding
      her that decided to cooperate and all those subsequent players in the queue.

© Springer-Verlag GmbH Germany, part of Springer Nature 2018
N. T. Nguyen et al. (Eds.): TCCI XXXI, LNCS 11290, pp. 1–8, 2018.
https://doi.org/10.1007/978-3-662-58464-4_1

That is, the player in her turn has the choice to cooperate or not to do it. The first choice is rewarded to her by her gain capacity in the game, while in the second she is rewarded by her blocking capacity.

Thus, not only the $n!$ orderings of a queue need to be considered, also the two choices for players need to be considered. This brings us to a more general model of $n! \cdot 2^n$ equally likely queues versus binary choices for the n players. The probabilistic model associated to this procedure is the discrete uniform distribution for all the $n! \cdot 2^n$ roll-calls, i.e., pairs formed by a permutation and a vector for players which determines if each player decides to cooperate or not.

In Bernardi and Freixas (2018) we already proposed this new formulation and did a detailed analysis for it. The main proof of the coincidence of our formula with the well-known formula by Shapley (1953) was proven by using power series and generating functions. This allowed us to give a succinct but perhaps not very intuitive prove. The main purpose of this work is to prove it with a more intuitive procedure based on a simpler technique: induction.

The new explicit formula for the Shapley value is of great interest. The main reason is its capability to be extended to more general contexts than cooperative or simple games. Felsenthal and Machover (1997) naturally extended simple games to ternary games, i.e., simple games in which voters are allowed to abstain as an intermediate option to vote favorably or against the issue at hand. They defined for this class of games a new value, with the same flavor of the Shapley value, based on what they call the ternary space of roll-calls. In terms of Freixas and Zwicker (2003), (2009) ternary games are a particular case of $(3, 2)$-simple games, which extend to $(j, 2)$-simple games, i.e., games in which players can choose among several ordered levels of approval and the output is binary. The natural extension of $(j, 2)$-simple games to the TU cooperative context is that of j-cooperative games, considered in Freixas (2018) in which players choose among different levels of activity and the characteristic function is defined on partitions capturing the choices of the players. The class of j-cooperative games is a bit more general than that of multichoice cooperative games, see Hsiao and Raghavan (1993).

The work in Freixas (2018) extends the formula we propose in this paper to j-cooperative games, and more particularly to: multichoice cooperative games, $(j, 2)$-simple games, and ternary games. Of course, when $j = 2$ the formula reduces to the formula (3) we propose in this paper.

But, as far as we know, it does not exist any explicit formula in these further contexts by using the original Shapley probabilistic scheme based only on queues and assuming that in her turn each player agrees to cooperate with her predecessors.

The setup of the rest of the paper is as follows. Section 2 includes some necessary preliminaries on the Shapley value. Section 3 is devoted to demonstrate the main result. The conclusions end the work in Sect. 4.

2 Preliminaries

Let N be a finite fixed set of cardinality n. The elements of N are called *players*, while a subset of N is called *coalition*. A *TU-cooperative game* is a function $v : 2^N \to \mathbb{R}$ such that $v(\emptyset) = 0$. The cardinality of a coalition S is denoted by s. Let $\mathcal{CG}_N$ be the set of all TU-cooperative games on N.

The *Shapley value* is a function $\phi : \mathcal{CG}_N \to \mathbb{R}^n$, that assigns to each player a real number $\phi_a(v)$. Shapley (1953), (1962) defined this function following a deductive procedure, by showing that it is uniquely characterized by the axioms of: efficiency, null player, symmetry and additivity.

The Shapley value has an explicit expression, in terms of the marginal contributions of the characteristic function, which is widely used to compute it:

$$\phi_a(v) = \sum_{S \subseteq (N \setminus \{a\})} \rho^n(s)[v(S \cup \{a\}) - v(S)], \tag{1}$$

where $s = |S|$ and

$$\rho^n(s) = \frac{s!(n-s-1)!}{n!}. \tag{2}$$

The coefficient $\rho^n(s)$, for each coalition $S \subseteq (N \setminus \{a\})$, is the proportion of permutations in which player a is occupying the $(s+1)$-position in the queue, where the preceding players are those in coalition S, no matter in which ordering, and the remaining players are occupying positions after $s+1$ in the queue, no matter in which ordering. Figures 1 and 2 schematically represents it.

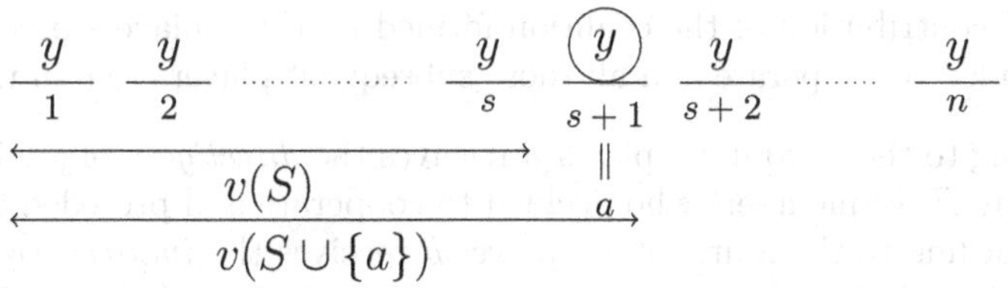

Fig. 1. Standard scheme for the classical Shapley value: "y" means that players choose forming part of the coalition.

The marginal contribution of a is weighted by a coefficient that counts all possible orderings of players before and after a as schematically described in Fig. 2.

3 The Alternative Explicit Formula for the Shapley Value

The alternative explicit formula following the more detailed model of roll-calls takes into account that in her turn, player a can decide either to cooperate or not.

$$\Phi_a(v) = \sum_{S \subseteq (N \setminus \{a\})} \Gamma^n(s)[v(S \cup \{a\}) - v(S)], \tag{3}$$

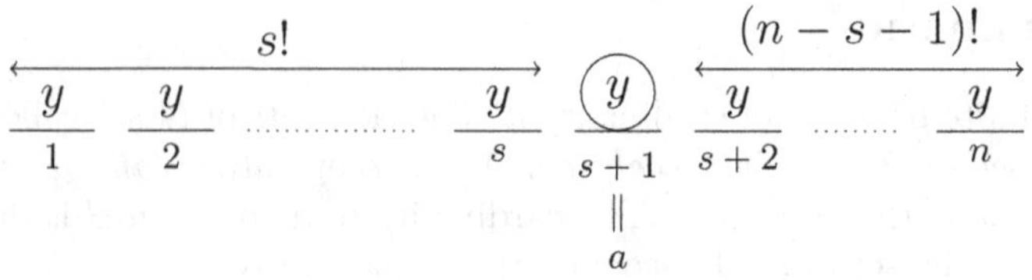

Fig. 2. Standard scheme for the classical Shapley value: counting all orderings.

where $s = |S|$ and for any $s = 0, \ldots, n-1$:

$$\Gamma^n(s) = \frac{s!}{2^n n!} \sum_{k=0}^{s} \frac{(n-k-1)!}{(s-k)!} 2^k + \frac{(n-s-1)!}{2^n n!} \sum_{k=0}^{n-s-1} \frac{(n-k-1)!}{(n-s-1-k)!} 2^k. \quad (4)$$

As we will see later Φ and ϕ coincide. The new formula is based on the following assumptions (see Theorem 4 in Bernardi and Freixas (2018)):

1. Players act in a randomly chosen order and all $n!$ orderings are equally likely.
2. In her turn, each player decides whether to cooperate or not in forming a coalition by either gaining collective value of blocking collective gain.
3. If in her turn the player has decided to cooperate, then she receives the marginal contribution of the coalition formed by those players preceding her that decided to cooperate.
4. If in her turn the player has decided not to cooperate, then she receives the marginal contribution of the coalition formed by those players preceding her that decided to cooperate and all those subsequent players to a in the queue.

According to the third item, player a receives the *direct gain* of joining to the coalition (say R) of members who decided to cooperate and preceded her in the queue. According to the fourth item, player a receives the *indirect gain*, i.e. the gain due to her blocking capacity, of joining to the coalition (say T) of members who decided to cooperate and preceded her in the queue (those in R) and all those players who follow her in the queue. The reason for the addition of the subsequent players to a in the queue is because we are assuming that in their turn they will decide to cooperate, which is the worst scenario for the capacity of player a to block gain.

Note that the expression in (3) is referenced to an arbitrary coalition S that does not contain player a as well as for the coefficients in (4). Thus, we need to consider $R = S$ when computing the direct gain, while for the indirect gain S is then the union of those who decided to cooperate and preceded player a in the queue and all those players who follow her in the queue. See Figs. 3 and 4.

Theorem 1. *The values Φ and ϕ for TU-cooperative games coincide.*

Proof. To prove Theorem 1 it is enough to deduce the equality of the coefficients $\rho^n(s)$ and $\Gamma^n(s)$ for all positive integers n and $0 \leq s \leq n-1$. Thus we need to

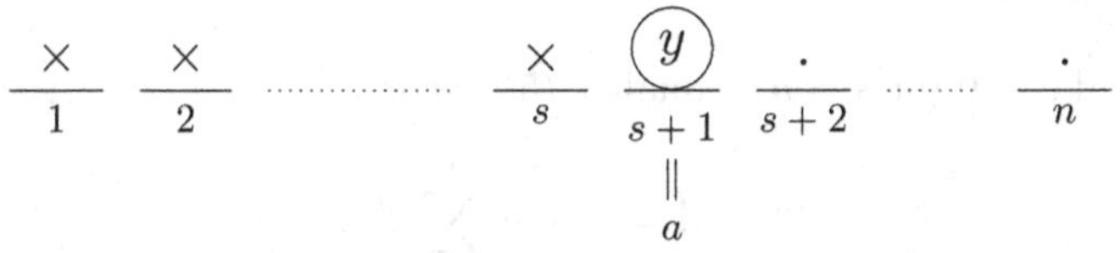

$$
\begin{array}{cccccccc}
\times & \times & \cdots\cdots & \times & \boxed{y} & \cdot & \cdots\cdots & \cdot \\
\hline
1 & 2 & & s & s+1 & s+2 & & n
\end{array}
$$

$$
\begin{array}{c}
\| \\
a
\end{array}
$$

Fig. 3. Scheme for direct gain: "×" means that players preceding a already decided to cooperate (y) or not (n). No matter if players after a are going to cooperate or not, which is represented by ".". Thus, coalition S is formed by those with $\times = y$.

$$
\begin{array}{cccccccc}
\times & \times & \cdots\cdots & \times & \boxed{n} & y & \cdots\cdots & y \\
\hline
1 & 2 & & s & s+1 & s+2 & & n
\end{array}
$$

$$
\begin{array}{c}
\| \\
a
\end{array}
$$

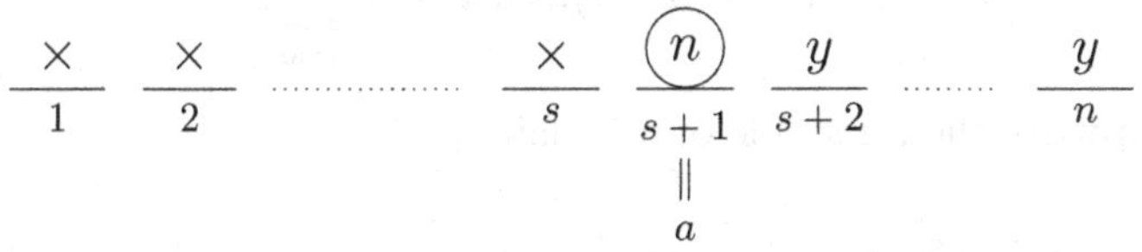

Fig. 4. Scheme for indirect gain: "×" means that players preceding a already decided to cooperate (y) or not (n). For measuring a's blocking gain capacity it is needed to assume that all players after a will choose to cooperate "y". Thus, coalition S is formed by those with $\times = y$ union all those after a in the queue.

prove that for any n and any $s = 0, \ldots, n-1$, it holds

$$
\frac{s!(n-s-1)!}{n!} = \frac{s!}{2^n n!}\sum_{k=0}^{s}\frac{(n-k-1)!}{(s-k)!}2^k + \frac{(n-s-1)!}{2^n n!}\sum_{k=0}^{n-s-1}\frac{(n-k-1)!}{(n-s-1-k)!}2^k.
$$

By simplifying the identity, the previous equality is equivalent to the following equation

$$
s!(n-s-1)!2^n = s!\sum_{k=0}^{s}2^k\frac{(n-k-1)!}{(s-k)!} + (n-s-1)!\sum_{k=0}^{n-s-1}2^k\frac{(n-k-1)!}{(n-s-k-1)!} \tag{5}
$$

for all n and any $s = 0, \ldots, n-1$.

Observe that if $n = 1$, then $s = 0$, then this equality reduces to $2 = 1 + 1$, that is trivially true. Since we are dealing with voting games, we assume that there is not only one player and so $n \geq 2$.

We proceed in proving (5) using induction on n.

If $n = 2$ and $s = 0$ we have $2^2 = 1 + 1 + 2$. If $n = 2$ and $s = 1$ we have $2^2 = 1 + 2 + 1$. So the thesis is true for $n = 2$.

Now, we assume that (5) is true for n and all $0 \leq s \leq n-1$ and we prove it for $n+1$ and $0 \leq s \leq n$.

We first consider the extreme cases $s = 0$ and $s = n$ and prove them directly. Secondly, we prove the statement for each s with $0 < s < n$, using the induction hypothesis for n with s and $s - 1$.

6   J. Freixas

First step:
For $n+1$ and $s=0$ (or $s=n$), equality (5) becomes

$$2^{n+1}n! = n! + n!\sum_{k=0}^{n} 2^k.$$

By the induction hypothesis (for n and $s=0$) we have

$$2^n(n-1)! = (n-1)! + (n-1)!\sum_{k=0}^{n-1} 2^k.$$

Then we can write the right side of our claim as

$$n! + n!\sum_{k=0}^{n} 2^k = n! + 2^n n! + n!\sum_{k=0}^{n-1} 2^k$$
$$= n! + 2^n n! + n[2^n(n-1)! - (n-1)!]$$
$$= n! + 2^n n! + 2^n n! - n! = 2^{n+1}n!$$

and this proves the first part.

Second step:
We now want to prove the thesis for $n+1$, thus, we have to show that the following is true

$$2^{n+1}s!(n-s)! \overset{?}{=} s!\sum_{k=0}^{s}\frac{(n-k)!}{(s-k)!}2^k + (n-s)!\sum_{k=0}^{n-1}\frac{(n-k)!}{(n-s-k)!}2^k. \quad (6)$$

By induction hypothesis, if we take n and s we have

$$s!(n-s-1)!2^n = s!\sum_{k=0}^{s}2^k\frac{(n-k-1)!}{(s-k)!} + (n-s-1)!\sum_{k=0}^{n-s-1}2^k\frac{(n-k-1)!}{(n-s-k-1)!} \quad (7)$$

and if we take $s-1$

$$(s-1)!(n-s)!2^n = (s-1)!\sum_{k=0}^{s-1}2^k\frac{(n-k-1)!}{(s-1-k)!} + (n-s)!\sum_{k=0}^{n-s}2^k\frac{(n-k-1)!}{(n-s-k)!}. \quad (8)$$

We work on the right-hand side of Eq. (6) and rewrite each of the two addends in the following way

$$s!\sum_{k=0}^{s}\frac{(n-k)!}{(s-k)!}2^k = s!(n-s)!2^s + s!\sum_{k=0}^{s-1}\frac{(n-k)!}{(s-k)!}2^k$$
$$= s!(n-s)!2^s + s!\sum_{k=0}^{s-1}\frac{(n-k-1)!}{(s-k-1)!}2^k\frac{n-k}{s-k},$$

writing $\frac{n-k}{s-k}$ as $\frac{n-s}{s-k}+1$,

$$= s!(n-s)!2^s + s!\sum_{k=0}^{s-1}\frac{(n-k-1)!}{(s-k-1)!}2^k\Big(\frac{n-s}{s-k}+1\Big)$$

$$= s!(n-s)!2^s + s!(n-s)\sum_{k=0}^{s-1}\frac{(n-k-1)!}{(s-k)!}2^k + s!\sum_{k=0}^{s-1}\frac{(n-k-1)!}{(s-k-1)!}2^k,$$

the first term can be moved inside the sum, to get

$$= s!(n-s)\sum_{k=0}^{s}\frac{(n-k-1)!}{(s-k)!}2^k + s!\sum_{k=0}^{s-1}\frac{(n-k-1)!}{(s-k-1)!}2^k.$$

Analogously the second term in (6) can be written as

$$(n-s)!\sum_{k=0}^{n-s}\frac{(n-k)!}{(n-s-k)!}2^k = s(n-s)!\sum_{k=0}^{n-s}\frac{(n-k-1)!}{(n-s-k)!}2^k + (n-s)!\sum_{k=0}^{n-s-1}\frac{(n-k-1)!}{(n-s-k-1)!}2^k.$$

If we now sum these expressions the right-hand side of (6) becomes

$$s\left[(s-1)!\sum_{k=0}^{s-1}\frac{(n-k-1)!}{(s-k-1)!}2^k + (n-s)!\sum_{k=0}^{n-s}\frac{(n-k-1)!}{(n-s-k)!}2^k\right]$$

$$+ (n-s)\left[s!\sum_{k=0}^{s}\frac{(n-k-1)!}{(s-k)!}2^k + (n-s-1)!\sum_{k=0}^{n-s-1}\frac{(n-k-1)!}{(n-s-k-1)!}2^k\right].$$

Using the induction hypothesis and in particular (7) and (8) and replacing everything in the right-hand side of (6), we finally get

$$2^{n+1}s!(n-s)! = s[2^n(s-1)!(n-s)!] + (n-s)[2^ns!(n-s-1)!]$$
$$= 2^ns!(n-s)! + 2^ns!(n-s)!$$
$$= 2^{n+1}s!(n-s)!$$

4 Conclusion

A new probabilistic approach to the Shapley value for TU-cooperative games has been proposed. Instead of just considering permutations as in the classical approach, the model also takes into account if the player wants to cooperate or not in her turn. From this new model we obtain a formula depending on the marginal contributions. This formula is proven by induction, which is a bit more transparent than using power series and generating functions as we recently did in Bernardi and Freixas (2018). This proof pretends to bring a little more light on different formulations of the Shapley value for TU-cooperative games.

Its main importance is revealed when trying extensions of the value to further contexts. Indeed, a natural extension of the formula (3) to the so-called j-cooperative games, games where players can choose among different (say j) ordered levels of activity, has been obtained in Freixas (2018). Nevertheless, as far as we know, it does not exist any formula of the value on j-cooperative games inspired in the original probabilistic model by Shapley based on marginal contributions on permutations.

Thus, the application potential of the model we present is enormous. All the theory and applications of cooperative games based on the Shapley value can be studied for the more general model of j-cooperative games, which contains multi-choice games.

Acknowledgements. This author research was partially supported by funds from the Spanish Ministry of Economy and Competitiveness (MINECO) and from the European Union (FEDER funds) under grant MTM2015-66818-P (MINECO/FEDER).

The author is grateful for the comments of two reviewers who have helped to improve the work.

References

Bernardi, G., Freixas, J.: The Shapley value analyzed under the Felsenthal and Machover bargaining model. Public Choice **176**(3–4), 557–565 (2018)

Brams, S.J.: Game Theory and Politics. Free Press, New York (1975)

Freixas, J.: A value for j-cooperative games: some theoretical aspects and applications. In: Algaba, E., Fragnelli, V., Sánchez-Soriano, J. (eds.) Contributed Chapters on the Shapley value, pp. 1–35. Taylor & Francis Group (2018, submitted)

Felsenthal, D.S., Machover, M.: Ternary voting games. Int. J. Game Theory **26**, 335–351 (1997)

Freixas, J., Zwicker, W.S.: Weighted voting, abstention, and multiple levels of approval. Soc. Choice Welf. **21**, 399–431 (2003)

Freixas, J., Zwicker, W.S.: Anonymous yes-no voting with abstention and multiple levels of approval. Games Econ. Behav. **69**, 428–444 (2009)

Hsiao, C.R., Raghavan, T.E.S.: Shapley value for multichoice cooperative games I. Games Econ. Behav. **5**, 240–256 (1993)

Luce, R.D., Raiffa, H.: Games and Decisions: Introduction and Critical Survey. Wiley, New York (1957)

Shapley, L.S.: A Value for N-person Games. In: Tucker, A.W., Kuhn, H.W. (eds.) Contributions to the Theory of Games II, pp. 307–317. Princeton University Press, Princeton (1953)

Shapley, L.S.: Simple games: an outline of the descriptive theory. Behav. Sci. **7**, 59–66 (1962)

Reflections on Two Old Condorcet Extensions

Hannu Nurmi[✉]

Department of Contemporary History, Philosophy and Political Science,
University of Turku, Turku, Finland
hnurmi@utu.fi

1 The Main Points

The distinction – in fact, rivalry – between two intuitive notions about what constitutes the winning candidates or policy alternatives has been present in the social choice literature from its Golden Age, i.e. in the late 18'th century [13]. According to one of them, the winners can be distinguished by looking a the performance of candidates in one-on-one, that is, pairwise contests. According to the other, the winners are in general best situated in the evaluators' rankings over all candidates. The best known class of rules among those conforming to the first intuitive notion are those that always elect the Condorcet winner whenever one exists. These rules are called Condorcet extensions for the obvious reason that they extend Condorcet's well-known winner criterion beyond the domain where it can be directly applied. A candidate is a Condorcet winner whenever it defeats all other candidates in pairwise contests with a majority of votes. [1]Condorcet extensions specify winners in all settings including those where a Condorcet winner is not to be found. Of course, in those settings where there is a Condorcet winner they all end up with electing it.

The focus of this article is on Condorcet extensions, more specifically on two choice rules dating back to the 19'th century, viz. Dodgson's and Nanson's procedures. The main points made here are:

- Condorcet winners are *prima facie* plausible solutions,
- they are in general incompatible with positional solutions,
- most of the Condorcet extensions can be given distance-minimizing representations,
- Dodgson's rule is such a rule,
- it is in a way self-contradictory,
- Nanson's rule is different from Borda elimination.

Useful remarks of the referees on an earlier version are gratefully acknowledged.

[1] Several variations of the criterion can be envisioned. To wit, one could call a candidate the Condorcet winner whenever he/she wins all contestants with more than 50% of the votes, or by some other qualified majority of votes. One can also determine the winner is pairwise contests by some other criterion like, for instance, the number of goals in football or time spent in skating a fixed distance, etc.

N. T. Nguyen et al. (Eds.): TCCI XXXI, LNCS 11290, pp. 9–21, 2018.
https://doi.org/10.1007/978-3-662-58464-4_2

The article is organized as follows. First we recall some basic concepts. Then we discuss briefly the plausibility of the Condorcet winner as the choice desideratum. Thereafter we focus on Dodgson's method and recall some of its bizarre properties. The next section deals with Nanson's rule and its main modifications. The final section draws some general conclusions.

2   Basic Concepts

We consider voting situations that consist of electorates having opinions about the candidates or policy alternatives. Let the generic electorate be a set N of n voters. Each voter is assumed to have a complete and transitive preference relation over the set A of k candidates.[2] A voting procedure or choice rule assigns a set of (winning) candidates to each $n+1$-tuple consisting of the set of candidates and n preference rankings over the candidates. The set of preference rankings is called a preference profile. This is the standard setting of the voting theory.

As mentioned above, the Condorcet winner is a candidate that defeats all its competitors in pairwise contests on the basis of the preferences of voters. In other words, x is the Condorcet winner if for any other candidate y, the number of voters ranking x higher than y is strictly larger than the number of those ranking y higher than x. Obviously, not all preference profiles contain a Condorcet winner.

Some other basic concepts are briefly characterized as follows:

- Condorcet loser: a candidate that would be defeated by all others in pairwise majority contests
- absolute winner (loser): a candidate ranked first(last) by an absolute majority of voters
- Dodgson's rule: given a profile of rankings over a set of candidates, count the minimum number of pairwise preference inversions between adjacent candidates that are needed to make each alternative the Condorcet winner
- Nanson's rule: given a profile of rankings, compute the Borda scores of each candidate. Then remove all those candidates with the average or smaller Borda score. Construct a new profile consisting of only the remaining candidates and repeat the elimination until the winner is found
- Borda elimination: first step as in Nanson's rule, but only the alternative with the smallest Borda score is eliminated.

3   The Plausibility of the Condorcet Winners

The Condorcet winner is, indeed, a very popular solution concept in voting games. Its appeal is evident in the various incompatibility results that have

---

[2] The descriptive accuracy of assuming complete and transitive preference relations can be and has been questioned, but in the present context we shall not dwell on the issue (see, e.g. [20]).

Table 1. Positionally dominated Condorcet winner

1 voter	1 voter	1 voter	1 voter	1 voter
D	E	C	D	E
E	A	D	E	B
A	C	E	B	A
B	B	A	C	D
C	D	B	A	C

appeared in scholarly outlets over the past decades (see e.g. [14,21]. Yet, its intuitive plausibility can be questioned (see esp. [22]). A serious objection against its general applicability was raised by Fishburn in terms of another plausible criterion of choice particularly suitable for use in preference profile contexts, viz. the positional dominance [9]. Table 1 illustrates. Here D is the Condorcet winner, but arguably E - the Borda winner - is a more plausible choice. To wit, D and E are ranked first by equally many voters (2), E is ranked second by two voters, while D is ranked second by no voters, E is ranked third by one voter, but D is ranked third by none. No voter ranks E lower than third, while D is ranked last by one and next to last by one voter. I.e. the Condorcet winner D is positionally dominated by another alternative, viz. E.

Another, perhaps less compelling, argument against the general plausibility of the Condorcet winner is based on the observation that it is not necessary that the Condorcet winner is among the largest vote getters in one-person-one-vote elections. Indeed, it is easy to envision a preference profile where the Condorcet winner receives the least number of votes. Table 2 gives one such profile. It is an a way an extreme case where the Condorcet winner C is ranked first by no voters at all.

Suppose that the candidate ranked first is the best representative of the voter. Table 2 shows that the Condorcet winner may be the least representative candidate of all since he/she (hereafter he)is ranked first by no voter, while all other candidates are best representatives of at least some voters. This is another demonstration of the potential for discrepant outcomes satisfying positional and binary criteria.

Table 2. Condorcet winner has the smallest number of first ranks

4 voters	3 voters	2 voters
A	B	D
C	C	C
B	A	B
D	D	A

Further problems undermining the general plausibility of the Condorcet winner have been pointed out by Saari [22]. Nonetheless, many - perhaps most - social choice scholars regard Condorcet consistency as the primary criterion for choice rules. This is certainly the underlying motivation of the two rules we now turn to.

4   What We Know about Dodgson' Rule

Dodgson's rule was brought to the awareness of the emerging social choice community by Duncan Black who reprinted some pamphlets of C.L. Dodgson (a.k.a. Lewis Carroll) from the late 19'th century [2].[3] The rule is in an obvious way motivated by the aim to elect Condorcet winners whenever those exist and something maximally similar when they don't. In a nutshell the rule singles out the Condorcet winner when the latter exist and otherwise elects the alternative that is closest to being a Condorcet winner in the sense that it can be made the Condorcet winner after a minimum number of binary preference switches between two adjacent alternatives. So, the rule adopts a metric for measuring closeness of being a Condorcet winner. This metric, generally known as the inversion metric, is one of many possibilities, but intuitively plausible as it stands.

Although, to the best of the present writer's knowledge, Dodgson's rule is not in practical use in any voting body, it is of great interest just because of its specific way of extending the notion of the Condorcet winner. This is what is generally known about the rule.

4.1   Condorcet Winner Choice

Dodgson's rule a Condorcet extension. This follows trivially from its definition since no other alternative can require less binary preference inversions than the Condorcet winner to become one.

4.2   Pareto Optimality

It is Pareto optimal, i.e. if candidate A Pareto-dominates candidate B, then B cannot become the Condorcet winner with fewer preference switches than A. To wit, all the switches needed to make A the Condorcet winner are needed to make B one as well and, moreover, to beat A, B needs a number of preference switches that is at least equal to half of the electorate, while A needs no switches at all to beat B. Hence, A's Dodgson score (the number of switches needed to become the Condorcet winner) is always strictly smaller than B's.

---

[3] Fishburn makes the valid point that it is not entirely fair to Dodgson to name the rule after him since he suggested several other rules and suggested counting preference switches only as a component of a more complex procedure [10]. Also Tideman questions the plausibility of associating Dodgson with this rule [24]. See [3]. Keeping these caveats in mind we shall, however, conform to the standard usage of the concept of Dodgson's rule.

Table 3. Nonmonotonicity of Dodgson's rule

42 voters	26 voters	21 voters	11 voters
B	A	E	E
A	E	D	A
C	C	B	B
D	B	A	D
E	D	C	C

4.3 Non-monotonicity

This was demonstrated by Fishburn [10] (see also [11]). The following example (Table 3) is from [19, p. 38].

Alternative A wins with only 14 binary preference reversals to become the Condorcet winner. Now, suppose that the 11 right-most voters increase the support of A by ranking it first, *ceteris paribus*. After the change, B is immediately below E in the 11-voter ranking and B needs only 9 binary preference changes to become the Condorcet winner, while A still needs 14. Therefore, the new winner is B. Thus, additional support for the winner (A), *ceteris paribus*, renders it a non-winner. Thus, Dodgson's rule is nonmonotonic.

4.4 Condorcet Loser Choice

That a Condorcet loser may require the least number of preference switches to become the Condorcet winner may strike as odd, but this can, indeed be the case. A weak version of this paradoxical occurrence was shown in [17, Ex. 5.4] where the Condorcet loser with some other alternatives receives the same minimal Dodgson score. A stronger case is the choice of an absolute loser, i.e. an alternative ranked last by more than half of the electorate. Clearly an absolute loser is a special case of a Condorcet loser in the sense that all absolute losers are Condorcet losers, but the converse is not true.

4.5 Absolute Loser Choice

In Table 4 candidate D needs only 3 preference inversions to become the Condorcet winner. All others need strictly more. Yet, with 15 last place rankings D is the absolute loser. Hence, Dodgson's rule may end up with an absolute loser (see also [19, p. 10]).

4.6 Computionally Intractability

The computation of the Dodgson winner is intuitively tedious in those situations where there is not Condorcet winner. After all, each alternative may be rendered the Condorcet winner through binary preference switches in several different

Table 4. Dodgson may elect an absolute loser

10 voters	7 voters	1 voter	7 voters	4 voters
D	B	B	C	D
A	C	A	A	C
B	A	C	B	A
C	D	D	D	B

ways. Of course only those involving the minimum number of switches count. The intuitive impression is corroborated by the results on computational complexity of determining the Dodgson winner. The first insights stem from [1] where it is shown that the determination of the Dodgson winner is NP-hard. Given C.L. Dodgson's widely known enthusiasm about puzzles, riddles and paradoxes, he would very likely have been thrilled to learn that the system bearing his name might in some circumstances involving large numbers of alternatives and voters take an inordinate amount of time and/or computational resource to yield a winner. This complexity is of course an aspect that speaks against the adoption of Dodgson's rule in large-scale elections.

The computational aspects of Dodgson's rule have positive implications as well, viz. manipulation of the rule is computationally intractable. This means that determining an effective way of preference misrepresentation may be computationally intractable [4]. Dodgson's rule is, thus, a mixed bag of good and bad properties when it comes to computational results: the complexity of preference misrepresentation certainly looks like a nice property, while the complexity in determining the winner is a bad one. The observations have, however, to be read in their proper context. They are the worst-case results that may have very little to do with the real world applications. Thus, for example, the computation of the Dodgson winner is quite tractable in profiles with a strong Condorcet or absolute winner. Similarly, if one's first ranked alternative is very low in other voters' preference rankings while his second-ranked alternative enjoys a widespread support as the first alternative with its toughest competitor being low in one's preference order, it does not require complex computations to conclude that one should switch the order between one's first and second ranked alternatives to contribute towards the victory of a pretty good, if not the best, alternative.

4.7 Variable-Sized Electorates

The well-known result of Moulin states that when the number of alternatives is strictly larger than three, all Condorcet extensions are vulnerable to the no-show paradox [14], i.e. for any Condorcet extension there are such profiles where a group of identically minded voters is better off by not voting at all than by voting according to their preferences. Since Dodgson's rule is a Condorcet extension, it is *eo ipso* vulnerable to the no-show paradox. The strong version of the no-show paradox occurs whenever the voting outcome changes from x to y when a group

Table 5. Dodgson's rule and the P-TOP paradox [5, p. 55]

42 voters	26 voters	21 voters	11 voters
B	A	E	E
A	E	D	A
C	C	B	B
D	B	A	D
E	D	C	C

of identically minded voters each preferring y to all other candidates chooses to abstain, *ceteris paribus*, instead of voting according to their preferences. In other words, abstaining leads to the best possible outcome for the voters not showing up at the polls. The strong version is sometimes called the P-TOP paradox [7,8].

Another type of no-show paradox occurs whenever a group of identically minded voters brings about their worst outcome by voting, while *ceteris paribus* their abstaining would result in a more preferable (from the group's point of view) outcome. This version is known as the P-BOT paradox [7,8]. It turns out that Dodgson's rule is vulnerable to both versions of the no-show paradox. Tables 5 and 6 demonstrate the vulnerability of Dodgson's rule to these extreme forms of the no-show paradox.

In Table 5 we have a 100-voter profile where there is no Condorcet winner, but A is the Dodgson winner, i.e. A requires fewer binary preference switches than the other candidates to become the Condorcet winner. Suppose now that 10 new voters with the preference ranking $A \succ B \succ E \succ C \succ D$ join the electorate while the remaining profile remains the same. I.e. the Dodgson winner A is at the top of the preference ranking of the new entrants. In the ensuing profile B becomes the Dodgson winner. Thus, we observe an instance of the P-TOP paradox.

Table 6 demonstrates the vulnerability of Dodgson's rule to the P-BOT paradox. Candidate B is the first ranked one in an absolute majority of the ballots making it the strong Condorcet or absolute winner. Hence, it requires no preference switches at all to become one and is, therefore, the Dodgson winner. Suppose now that a group of three like-minded voters with $A \succ D \succ B \succ C$ ranking join the electorate, *ceteris paribus*. In the ensuing profile of 12 voters, there is no Condorcet winner and candidate C, i.e. the lowest ranked candidates in the new entrants' preferences, becomes the Dodgson winner.

Apart from its successes and failures with respect to various social choice desiderata, Dodgson's rule is instructive in a more general sense, viz. it points to a peculiarity in the Condorcet winner criterion itself. To wit, Dodgson's rule is perhaps the most obvious extension of Condorcet's winning intuition to settings where a Condorcet winner does not exist. Its fundamental idea is to look for an alternative that is as close as possible to being a Condorcet winner, closeness being defined by the number of binary preference switches between adjacent candidates in the voters' preference rankings. So, both the "goal state" (where a

Table 6. Dodgson'r rule and the P-BOT paradox [5, p. 57]

5 voters	4 voters
B	C
C	D
D	A
A	B

Condorcet winner exists) and the distance between the observed preference profile and the goal state are of binary rather than positional nature. And yet, the alternative which is defeated by all others in pairwise comparisons with a majority of votes (Condorcet loser) may become the Dodgson winner. In fact, even the strong Condorcet loser or the absolute loser may be elected by Dodgson's rule (see Table 4 above).

We now turn to another choice of roughly the same period, viz. the end of the 19'th century: Nanson's rule.

5 Three Versions of Nanson's Rule

In contrast to Dodgson, Nanson was well aware of the results of the earlier social choice scholars, especially of Borda and Condorcet. He was primarily interested in combining the two winning intuitions: the binary one underlying Condorcet's method and the positional one exhibited by the Borda count [15]. Like Dodgson, Nanson wanted to secure the election of the Condorcet winner whenever it exists. However, as there often is no Condorcet winner, he suggested using Borda's intuition of winning. Similar idea was some seven decades later adopted by Black in a more straight-forward manner as a consecutive application of the Condorcet's winning principle and then - should it not be applicable - Borda's method of marks [2]. Nanson's idea was to use a single principle and yet come up with outcomes that combine what he saw as two main virtues of Borda's and Condorcet's methods: the majority support of the winner (Condorcet) and decisiveness (Borda). The principle introduced is the use of Borda count as a method of successively eliminating candidates on the basis of their (poor) showing in order to finally end up with the winner as the sole non-eliminated candidate.

How, then, should one proceed in the elimination? Nanson's [15, p. 211] view is unambiguous:

"Proceed exactly as in Borda's method, but instead of electing the highest candidate, reject all who have not more than the average number of votes polled. If (in the case of three candidates) two be thus rejected, the election is finished; but if one only is rejected, hold a final election between the two remaining candidates on the usual plan."

Table 7. Different choice by BER and NER

8 voters	5 voters	5 voters	2 voters
A	C	B	C
B	A	C	A
C	B	A	B

In modern literature Nanson's rule appears in three different senses:

- Borda elimination (BER): eliminate the candidate(s) with the smallest Borda score, then repeat the procedure for the remaining candidates until one candidate only (the winner) remains
- Schwartz's elimination rule (SER): "Reckon the Borda count for all the feasible alternatives and eliminate those whose Borda counts are less than the mean Borda count, then do the same to the reduced feasible set, and continue this way until further eliminations are impossible." [23, p. 180]
- Nanson's true rule (NER) defined above.

BER appears in Fishburn's path-breaking article under the name "Nanson's function" [10, p. 473], not because of the author's mistake concerning Nanson's rule, but as a matter of convention (see [16, p. 191] and [12]). The second variation (SER) has not been used in the literature. So, we focus mainly on BER and NER.

The differences between BER and NER are exhibited already in three-candidate contests, as shown in Table 7.

In Table 7 BER eliminates first B, and then elects C. NER elects A as the sole candidate with strictly larger than average Borda score, viz. 23.

The fact that the choices of the two rules can be different, does not mean that this is often or generally, much less always, the case. More importantly, there are properties that both rules share. The most significant of these is undoubtedly the satisfaction of the Condorcet winner criterion. In fact, this was the main motivation of NER and can be readily verified by observing that there is a loose connection between the Condorcet winner and the Borda scores: the former can never have a very low Borda score; rather it has to have a strictly higher than average Borda score. Hence by eliminating all alternatives with at most the average Borda score one can rest assured that the eventual Condorcet winner survives the elimination process and will accordingly be elected. This property is retained in BER and SER.

The situation changes somewhat when we turn our attention to the Condorcet condition that pertains to profiles where there is a set of majority undominated alternatives, but no Condorcet winner. Sometimes such a set is called the core of the voting game. A condition variably called the weak Condorcet condition [16, p. 192] or the strict Condorcet principle [10, p. 479] says that all candidates in the core and only them ought to be elected. In terms of this criterion we observe some variation among the versions of Nanson's rule [18, pp. 201–203]. Fishburn shows that BER violates this principle [10] and Schwartz

Table 8. NER and core distinct candidates

1 voter	1 voter	1 voter	1 voter	2 voters
A	A	A	B	C
B	D	D	C	E
C	B	E	E	D
D	C	B	D	B
E	E	C	A	A

makes the same observation regarding SER [23, p. 181]. Niou provides an example where also NER fails on this principle [16, pp. 192–193]. These examples do not, however, tell the whole story. To wit, the violations reported above pertain to cases where the BER, SER and NER choice sets include the core but, in addition, some other candidates not in the core. The variation alluded to above stems from the fact that NER can elect an alternative that is totally distinct from the core. This is demonstrated in Table 8 [18, p. 202]. In this profile the average Borda score is 12. Only candidate C has a Borda score that exceeds this. Hence the NER winner is C. However, there is a nonempty core that consists of A alone.[4] It is evident that BER and SER are not vulnerable to this, more dramatic failure on the strict Condorcet principle for a core alternative cannot have a strictly smaller than average Borda score. Thus, there is potential price to be paid for swapping BER or SER to NER, whereas the latter is no doubt more efficient in finding the winner.

There are other minor differences as well. To wit, it can be shown that with three candidates, BER is nonmonotonic [10, p. 478], while NER isn't [18, pp. 204–205]. BER's nonomotonicity is demonstrated by Table 9. There BER elects A. Add now 2 voters with $A \succ C \succ B$ ranking. Then C ends up the winner, demonstrating BER's vulnerability to P-TOP. N.B. NER elects A as well initially, but elects it also in the augmented electorate. In larger candidate sets, however, all Nanson variations are nonmonotonic.

The relationships between BER and NER with respect to monotonicity violations can be summarized as follows [6].

Result 1. *Felsenthal and Nurmi 2017. Given more than three candidates and in the absence of a Condorcet winner, BER and NER are both vulnerable to monotonicity violations in both fixed and variable electorates; moreover, profiles exist such that BER is vulnerable to monotonicity failure but NER is not as well as vice versa and regardless of whether the initial winner according to BER is the same as, or different than, that according to NER.*

In other words BER and NER are largely independent of each other with regard to monotonicity violations.

---

[4] Note that the NER winner C is defeated by B in a pairwise majority comparison. Hence C is not in the core.

Table 9. P-TOP and BER (but not NER)

5 voters	4 voters	4 voters
A	B	C
B	C	A
C	A	B

6 By Way of a Conlusion

So, what have we learned? What is the point in studying old voting procedures that are used nowhere? The main lesson pertains to voting system design. All voting systems - old and new - are based on a more or less well-articulated notion of what constitutes a winner or the best choice given a profile of preferences (or a set of performance rankings with respect to properties that are deemed important in an MCDM context). Once a procedure is set up in accordance with this notion, some properties emerge that are found to be undesirable. Hence, the procedure is modified to accommodate the problems. Yet, when fixing one problem one often creates another one, a problem that the original procedure did not have. This pattern is exemplified by the above two old Condorcet extensions. Their starting point must have been a conviction that the Condorcet winner ought be elected whenever available. Since there may not always be a Condorcet winner, Dodgson and Nanson took two separate routes to finding a plausible winner. Dodgson looked for a candidate that would be as close as possible to being a Condorcet winner. The closeness measure used is very much in the spirit of pairwise comparisons, viz. how many binary switches of candidates one needs to make in the preference profile in order to render a candidate the Condorcet winner? The fewer, the closer. Nanson, in turn, tried to combine both binary and positional intuitions of winning by using the Borda count in an elimination process so designed that the Condorcet winner could not be eliminated. In fact Nanson used a single principle - not a hybrid of two - in choosing the winner: the elimination of candidates with at most the average Borda score. It is a derived result of such a process that the eventual Condorcet winner will not be eliminated and will thus be elected.

Both Dodgson's and Nanson's rules thus solve the problem of electing the Condorcet winner when one exists. The way the solution is devised is, however, accompanied with major flaws: their nonmonotonicity both in fixed and variable electorates. This is particularly significant in Nanson's rule since the procedure used in the elimination process, the Borda count, is monotonic. So, by fixing one problem of the Borda count, viz. the occasional non-election of the Condorcet winner, Nanson created another one that does afflict the Borda count.

In the case of Dodgson's rule the distance measurement confronts us with a bizarre possibility: the absolute loser and *a fortiori* the Condorcet loser may be the closest to being the Condorcet winner not only in the sense of belonging to a set of tied alternatives, but in the sense of being the unique candidate that is closest to being the Condorcet winner.

The main lesson from this brief exercise is that modifying systems often involves trade-offs and the prudent way to proceed is to examine the entire set of properties that are significant in the choice problems at hand. The significance often depends on the types of profiles one expects to encounter: how many candidates are typically in the race, how large is the typical electorate, is the typical electorate homogeneous or divided into many opposing groups and so on.

References

1. Bartholdi, J., Tovey, C., Trick, M.: Voting schemes for which it can be difficult to tell who won the election. Soc. Choice Welf. **6**, 157–165 (1989)
2. Black, D.: The Theory of Committees and Elections. Cambridge University Press, Cambridge (1958)
3. Brandt, F.: Some remarks on Dodgson's voting rule. Math. Log. Q. **55**, 460–463 (2009)
4. Caragiannis, I., Hemaspaandra, E., Hemaspaandra, L.: Dodgson's rule and Young's rule. In: Brandt, F., Conitzer, V., Endriss, U., Lang, J., Procaccia, A. (eds.) Handbook of Computational Social Choice, pp. 103–126. Cambridge University Press, New York (2016)
5. Felsenthal, D.S., Nurmi, H.: Monotonicity Failures Afflicting Procedures for Electing a Single Candidate. SE. Springer, Cham (2017). https://doi.org/10.1007/978-3-319-51061-3
6. Felsenthal, D.S., Nurmi, H.: Monotonicity violations by Borda elimination and Nanson's rules: a comparison Group Decis. Negot. **27**, 637–664 (2018)
7. Felsenthal, D.S., Tideman, N.: Varieties of failure of monotonicity and participation under five voting methods. Theory Decis. **75**, 59–77 (2013)
8. Felsenthal, D.S., Tideman, N.: Interacting double monotonicity failure with direction of impact under five voting rules. Math. Soc. Sci. **67**, 57–66 (2014)
9. Fishburn, P.C.: The Theory of Social Choice. Princeton University Press, Princeton (1973)
10. Fishburn, P.C.: Condorcet social choice functions. SIAM J. Appl. Math. **33**, 469–489 (1977)
11. Fishburn, P.C.: Monotonicity paradoxes in the theory of elections. Discret. Appl. Math. **4**, 119–134 (1982)
12. Fishburn, P.C.: A note on "A note on Nanson's rule". Public Choice **64**, 101–102 (1990)
13. McLean, I., Urken, A. (eds.): Classics of Social Choice. The University of Michigan Press, Ann Arbor (1995)
14. Moulin, H.: Condorcet's principle implies the no-show paradox. J. Econ. Theory **45**, 53–64 (1988)
15. Nanson, E.J.: Methods of election. Trans. Proc. R. Soc. Vic., Art. XIX, 197–240 (1883)
16. Niou, E.M.S.: A note on Nanson's rule. Public Choice **54**, 191–193 (1987)
17. Nurmi, H.: Comparing Voting Systems. D. Reidel, Dordrecht (1987)
18. Nurmi, H.: On Nanson's method. In: Apunen, O., Borg, O., Hakovirta, H., Paastela, J. (eds.) Democracy in the Modern World. Essays for Tatu Vanhanen, Series A, vol. 260, pp. 199–210. Acta Universitatis Tamperensis, Tampere (1989)

19. Nurmi, H.: Monotonicity and its cognates in the theory of choice. Public Choice **121**, 25–49 (2004)
20. Nurmi, H.: Are we done with preference rankings? If we are, then what? Oper. Res. Decis. **24**, 63–74 (2014)
21. Pérez, J.: The strong no show paradoxes are a common flaw in Condorcet voting correspondences. Soc. Choice Welf. **18**, 601–616 (2001)
22. Saari, D.G.: Basic Geometry of Voting. Springer, Heidelberg (1995). https://doi.org/10.1007/978-3-642-57748-2
23. Schwartz, T.: The Logic of Collective Choice. Columbia University Press, New York (1986)
24. Tideman, N.: Independence of clones as a criterion for voting rules. Soc. Choice Welf. **4**, 185–206 (1987)

Transforming Games with Affinities from Characteristic into Normal Form

Cesarino Bertini[1], Cristina Bonzi[2], Gianfranco Gambarelli[1],
Nicola Gnocchi[3], Ignazio Panades[1], and Izabella Stach[4(✉)]

[1] Department of Management, Economics and Quantitative Methods,
University of Bergamo, Via dei Caniana, 2, 24127 Bergamo, Italy
{cesarino.bertini,gianfranco.gambarelli}@unibg.it,
panades.i@hotmail.com
[2] TWIG srl, Bergamo, Italy
cristina.bonzi23@gmail.com
[3] Marketing Department, Nolan Group spa, Brembate di Sopra, Italy
n.gnocchi@nolan.it
[4] AGH University Science and Technology, Al. Mickiewicza 30,
30-059 Krakow, Poland
istach@zarz.agh.edu.pl

Abstract. von Neumann and Morgenstern, while introducing games in extensive form in their book [1], also supplied a method for transforming such games into normal form. Once more in their book [1], the same authors provided a method for transforming games from characteristic function form into normal form, although limited to constant-sum games. In his paper [2], Gambarelli proposed a generalization of this method to variable-sum games. In this generalization, the strategies are the requests made by players to join any coalition, with each player making the same request to all coalitions. Each player's payment consists of the player's request multiplied by the probability that the player is part of a coalition really formed. Gambarelli introduced a solution for games in characteristic function form, made up of the set of Pareto-Optimal payoffs generated by Nash Equilibria of the transformed game.

In this paper, the above transformation method is generalized to the case in which each player's requests vary according to the coalition being addressed. Propositions regarding the existence of a solution are proved. Software for the automatic generation of the solution is supplied.

Keywords: Game theory · Characteristic function · Nash equilibria
Pareto optimality · Transformation

1 Introduction

While presenting games in extensive form in their book [1], von Neumann and Morgenstern also advanced a method for transforming such games into normal form, albeit with the loss of certain information from the original game, as illustrated, for example, by Robert Aumann in his book [3]. Once more in their book [1], the same authors provided a method for transforming games in characteristic function form into normal

© Springer-Verlag GmbH Germany, part of Springer Nature 2018
N. T. Nguyen et al. (Eds.): TCCI XXXI, LNCS 11290, pp. 22–36, 2018.
https://doi.org/10.1007/978-3-662-58464-4_3

form, although limited to constant-sum games. In his paper [2], Gambarelli proposed a generalization of this method to variable-sum games. In this model, the strategies are the requests made by players to join any coalition, with each player making the same request to all coalitions. Each player's payment consists of his request multiplied by the probability that he/she is part of a coalition really formed. Gambarelli introduced a solution for games in characteristic function form, made up of the set of Pareto-optimal payoffs generated by Nash equilibria of the transformed game. We refer the reader to that paper for a review of classical literature on the matter.

In this paper, the above transformation method is generalized to the case in which each player's requests vary according to the coalition being addressed. Propositions regarding the existence of a solution, together with software for its related automatic computation, are supplied.

The paper is structured as follows. The preliminary definitions devoted to the game theoretical notion and notations used throughout the paper are given in Sect. 1.1. The new model is introduced in Sect. 2. Following this, in Sect. 3, we show the algorithm for generating the transformed game and its solution. A numerical example is presented in Sect. 4. Some propositions regarding the existence of a solution are given in Sect. 5. Section 6 is dedicated to conclusions and open problems. Software for the automatic search for the solution is supplied in the Appendix.

1.1 Preliminary Definitions and Notations

In this section we give the standard notation and definitions from game theory used throughout the paper.

Let $N = \{1, \ldots, n\}$ be a finite set, where n is an integer number and $n > 0$. From here onwards, unless otherwise specified, the index i is intended to represent the i-th player (element of N) and S represents a generic coalition (subset of N). By 2^N we denote the powerset of N. A *cooperative game in characteristic form* is a pair (N, v), where $v : 2^N \rightarrow R$ is characteristic function, with $v(\varnothing) = 0$. If N is fixed, we identify game (N, v) with its characteristic function v. Any $x = (x_1, \ldots, x_n) \in R^n$ is said to called a *payoff vector* or *payment*. A cooperative game is said to be *game with transferable utility* (TU game for short) if it is implicitly assumed that a coalition S can distribute its value $v(S)$ to its members in any way they choose. Payment $x = (x_1, x_2, \ldots, x_n)$ is called to be *feasible payment* if $\sum_{i \in N} x_i \leq N$. Payment $x = (x_1, x_2, \ldots, x_n)$ is called to be *Pareto optimal* if $\sum_{i \in N} x_i = N$, what means that no player can improve its payoff without lowering the payoff of another agent. Game v is said to be *simple game* if $v(S) \in \{0, 1\}$ for every $S \subseteq N$. Then, game v is said to be *monotonic* if $v(S) \leq v(T)$ holds for all $S, T \subseteq N$ such that $S \subseteq T$ and it is said to be *superadditive* if $v(S \cup T) \geq v(S) + v(T)$ holds for all disjunctive coalitions $S, T \subseteq N$. A game v is said to be *inessential* (or *additive*) if $v(S) = \sum_{i \in S} v(\{i\})$ for every $S \subseteq N$ and it is said to be *subadditive* if $v(S \cup T) \leq v(S) + v(T)$ holds for all disjunctive coalitions $S, T \subseteq N$. Moreover, game v is said to be *constant-sum game* if $v(S) + v(N \backslash S) = v(N)$ for every $S \subseteq N$.

Let $x = (x_1, x_2, \ldots, x_n)$ be a payment of the TU game (N, v). The *core* of a game v is set $C(v)$ defined as follows $C(v) = \{x \in R^n : \sum_{i \in N} x_i = v(N), \sum_{i \in S} x_i \geq v(S) \text{ for every } S \subseteq N\}$.

A *Nash equilibrium* is a set of strategies, one for each player, such that no player has incentive to change his or her strategy given what the other players are doing. A strategy profile $x = (x_1^*, x_2^*, \ldots, x_n^*)$ states a Nash equilibrium of n-person game if for each player $i \in N$ and for all x_i the following condition holds: $(x_1^*, \ldots, x_{i-1}^*, x_i^*, x_{i+1}^*, \ldots, x_n^*) \geq (x_1^*, \ldots, x_{i-1}^*, x_i, x_{i+1}^*, \ldots, x_n^*)$..

2 Model

In this section, we present the game transformation and the solution.

2.1 Data of Original Game

Let v be a non-negative TU game in characteristic function form, defined on the set of players $N = \{1, \ldots, n\}$.

Furthermore, for all coalitions S of N to which the i-th player belongs, the indices a_i^S are given, these expressing the *level of affinity* of each player for that particular coalition.

These indices are subject to $0 < a_i^S \leq 1$, with the obvious condition that the agreement of each player is in itself the maximum: $a_i^{\{i\}} = 1$.

The ratio $d_i^S = 1/a_i^S$ represents the *disagreement* of the i-th player for coalition S.

2.2 Strategies of Transformed Game

A *strategy profile* $x = \{x_i\}$ corresponds to the *basic request* for payment, made by players, under the conditions $v(\{i\}) \leq x_i \leq v(N)$. Every *effective request* x_i^S made by the i-th player to coalition S is the product of the player's basic request and the player's disagreement with that coalition: $x_i^S = x_i/a_i^S$.

It is, in fact, reasonable that a player makes a request on the basis of the degree of affinity, whether greater or smaller, with that particular coalition. How much should a player demands for taking part in a coalition? In case of perfect affinity disagreement reaches the minimum equal to 1. The larger the disagreement, the larger the request made by a player to join the coalition.

2.3 Payoffs and Solution of Transformed Game

In this paragraph we will report some concepts that are presented more specifically in Gambarelli [2]. We call $e^S(x)$ the probability that coalition S will be formed. We define the expected payoff $p_i^S(x)$ of the i-th player from S as follows: it is 0, if i does not belong to S, otherwise it is the product of his request to S and the probability that this coalition will be formed:

$$p_i^S(x) = \begin{cases} x_i^S e^S(x) & \text{if } i \in S \\ 0 & \text{elsewhere.} \end{cases}$$

The *payoff* $p_i(x)$ of the i-th player consists of the total of his expected payoffs from all coalitions:

$$p_i(x) = \sum_{S \subseteq N} p_i^S.$$

To reach a complete definition of the payoffs, the probability $e^S(x)$ that coalition S will be formed must be determined. We shall now do this.

We call *possible winning coalition* (PWC) any coalition S that is able to satisfy all requests from its members, i.e. such that:

$$\sum_{i \in S} x_i^S \leq v(S).$$

All other coalitions are called *losing coalitions*.

Having established a strategy profile x for every possible winning coalition S, we call *bunch* $B^S(x)$ the set of possible winning coalitions that have a non-empty intersection with S, in union with all possible winning coalitions that have a non-empty intersection with preceding ones, and so on, for all possible winning coalitions of N. It is easy to verify that these bunches form a partition of N in disjoint subsets.

For each bunch $B^S(x)$, we call *'winning layout'* $L(B^S(x))$ any set of coalitions of $B^S(x)$ that have a two by two empty intersection, and are such that the union of their players is the set of players of $B^S(x)$.

Having established a strategy profile x, the probability $e^S(x)$ that coalition S will be formed is defined, according to the classical concept, by the ratio between the number of layouts to which S belongs and the total number of layouts in its bunch.

We define the *solution of v via transformed game*, or simply the *TG-solution of v*, as the set of Pareto-optimal payoffs resulting from the Nash equilibria of the transformed game.

3   Algorithm

In this Section, we give a discrete algorithm for transforming the game and for computing the TG-solution. In the appendix we provide a listing of software program. So, it is possible to see, the logical sequence of steps that follow in solving a problem.

The first step of the algorithm refers to the acquisition of the input data. In this step, values $v(S)$ and affinities a_i^S are acquired and checked. Furthermore, the density of the grid where the algorithm must operate is acquired; that is, the measurement δ of the side of the hypercubes of this grid. The gird is constructed as follows. Let n be the number of players in game v and let $\delta > 0$ be stated. The gird for this game consists of points $(x_1, \ldots, x_n)$ such that $x_i = v(\{i\}) + k \ldots \delta, k = 0, 1, \ldots, \frac{v(N) - v(\{i\})}{\delta}$ for all $i = 1, \ldots, n$.

After the input data step, the computations start. Varying i from 1 to n, any basic request x_i is considered, in order to reach all points of the grid, from $v(\{i\})$ to $v(N)$. The requests to coalitions, and the payments, are calculated for each x_i, as indicated in Sects. 2.2 and 2.3. The outcome thus obtained is checked to see whether it is Pareto-optimal. If a payment is not Pareto-optimal, then the request is discarded and the next one considered. Otherwise, the request is verified for correspondence to a Nash equilibrium and, if so, it is printed as a solution.

For each point of the solution found, players' requests to the winning coalitions of which they are part are reported, together with consequent payoffs. In the event the algorithm generates no solution, this should be announced.

Considering technical remarks on the gird, of course, the smaller δ, the greater the precision required and, consequently, the greater the elaboration time needed. It may happen that, by decreasing δ, some solutions previously found will be lost as the points of investigation change. In any case, the discrete nature of the algorithm means that not all solutions may be achieved; for instance irrational ones.

4   Example

Data are given in Table 1. We restrict ourselves to giving the results of some calculations.

Table 1. Data of our example.

Coalition S	Characteristic function $v(S)$	Affinities		
		a_1	a_2	a_3
$\{1\}$	0	1	–	–
$\{2\}$	0	–	1	–
$\{3\}$	1	–	–	1
$\{1, 2\}$	1	1/2	1/3	–
$\{1, 3\}$	1	1/5	–	2/3
$\{2, 3\}$	2	–	1/3	1
$\{1, 2, 3\}$	3	1/5	1/3	2/3

4.1   First Step

- $x_1 = 0$, $x_2 = 0$, $x_3 = 1$.

Given the basic requests of the players, we compute the specific requests to the coalitions, using $x_i^S = \frac{x_i}{a_i^S}$. Table 2 reports the specific requests of every player.

Table 2. Specific requests at first step.

Player 1	$x_1^{\{1\}} = 0.00$	$x_1^{\{1,2\}} = 0.00$	$x_1^{\{1,3\}} = 0.00$	$x_1^{\{1,2,3\}} = 0.00$
Player 2	$x_2^{\{2\}} = 0.00$	$x_2^{\{1,2\}} = 0.00$	$x_2^{\{2,3\}} = 0.00$	$x_2^{\{1,2,3\}} = 0.00$
Player 3	$x_3^{\{3\}} = 1.00$	$x_3^{\{1,3\}} = 1.50$	$x_3^{\{2,3\}} = 1.00$	$x_3^{\{1,2,3\}} = 1.50$

Now, we have to find the *possible winning coalitions*. These coalitions are:

$\{1\}$ because $x_1^{\{1\}} = 0 \le v(\{1\}) = 0$

$\{2\}$ because $x_2^{\{2\}} = 0 \le v(\{2\}) = 0$

$\{3\}$ because $x_3^{\{3\}} = 1 \le v(\{3\}) = 1$

$\{1, 2\}$ because $x_1^{\{1,2\}} + x_2^{\{1,2\}} = 0 + 0 \le v(\{1, 2\}) = 1$

$\{2, 3\}$ because $x_2^{\{2,3\}} + x_3^{\{2,3\}} = 0 + 1 \le v(\{2, 3\}) = 2$

$\{1, 2, 3\}$ because $x_1^{\{1,2,3\}} + x_2^{\{1,2,3\}} + x_3^{\{1,2,3\}} = 0 + 0 + \frac{3}{2} \le v(\{1, 2, 3\}) = 3$.

In this case, we have only one non-empty coalition which is not possible wining coalition, it means coalition $\{1, 3\}$. Namely coalition $\{1, 3\}$ does not satisfy the condition $\sum_{i \in S} x_i^S \le v(S)$. More precisely $x_1^{\{1,3\}} + x_3^{\{1,3\}} = 0 + 1.5 > v(\{1, 3\}) = 1$.

We generate the bunches. In this case there is a single bunch:

$$\{\{1\}, \{2\}, \{3\}\}, \{\{1, 2\}, \{3\}\}, \{\{2, 3\}, \{1\}\}, \{\{1, 2, 3\}\}.$$

Now we have to compute the expected value for every player:

Player 1: $e_1^{\{1\}} = 1/2$, $e_1^{\{1,2\}} = 1/4$, $e_1^{\{1,2,3\}} = 1/4$

Player 2: $e_2^{\{2\}} = 1/4$, $e_2^{\{1,2\}} = 1/4$, $e_2^{\{2,3\}} = 1/4$, $e_2^{\{1,2,3\}} = 1/4$

Player 3: $e_3^{\{3\}} = 1/2$, $e_3^{\{2,3\}} = 1/4$, $e_3^{\{1,2,3\}} = 1/4$

So we can compute the payoffs according to the formula presented in Sect. 2.3:

Player 1: $0 \cdot \frac{1}{2} + 0 \cdot \frac{1}{4} + 0 \cdot \frac{1}{4} = 0$;

Player 2: $0 \cdot \frac{1}{4} + 0 \cdot \frac{1}{4} + 0 \cdot \frac{1}{4} + 0 \cdot \frac{1}{4} = 0$;

Player 3: $1 \cdot \frac{1}{2} + 1 \cdot \frac{1}{4} + 1.5 \cdot \frac{1}{4} = \frac{9}{8}$.

So: $p(x) = [0, 0, \frac{9}{8}]$.

The sum of the payoffs is below the band for Pareto-optimality, so we therefore discard this vector of strategies and move on to the next one.

4.2 Intermediate Step

In this section we report the computation regarding the step where the basic requests are $x_1 = 0.20$, $x_2 = 0.50$, $x_3 = 1.10$. This request can be obtain by the algorithm changing the gird point of investigation with $\delta = 0.1$. The specific requests of the players are reported in Table 3. It is easy to verify that this step does not supply winning coalition.

Table 3. Specific requests at the intermediate step.

PLAYER 1	$x_1^{\{1\}} = 0.20$	$x_1^{\{1,2\}} = 0.40$	$x_1^{\{1,3\}} = 1.00$	$x_1^{\{1,2,3\}} = 1.00$
PLAYER 2	$x_2^{\{2\}} = 0.50$	$x_2^{\{1,2\}} = 1.50$	$x_2^{\{2,3\}} = 1.50$	$x_2^{\{1,2,3\}} = 1.50$
PLAYER 3	$x_3^{\{3\}} = 1.10$	$x_3^{\{1,3\}} = 1.65$	$x_3^{\{2,3\}} = 1.10$	$x_3^{\{1,2,3\}} = 1.65$

4.3 Solution

The results of our example, using $\delta = 1/10$, are shown in Table 4. We can see that all the strategic profiles leading to the solutions have only one possible winning layout. Owing to this unique layout, the specific payoffs for each player are equal to the specific requests.

Table 4. Discrete solution to our example.

PWC	$v(S)$	Specific requests = specific payoffs		
		1	2	3
{2, 3}	2		0.90	1.10
{1, 2}	1	0.40	0.60	
{3}	1			1.00

5 On the Existence of Solution

In this section we provide some propositions regarding the existence of a solution.

Proposition 1. The TG-solution is not empty for all subadditive games.

Proof: It is easy to verify that the TG-solution of these games is $p_i = v(\{i\})$ for all $i \in N$. QED

Remark 1. Since subadditive games have an empty core, it is implicit that the TG-solution of these games does not belong to the core.

Proposition 2. The TG-solution is not empty for all inessential games.

Proof: It is easy to verify that, as in the case of subadditive games, the TG-solution of the inessential games is $p_i = v(\{i\})$ for all $i \in N$. QED

Remark 2. The TG-solution for inessential games coincides with their core.

Proposition 3. The TG-solution is not empty for all superadditive games, such that: a coalition R of N exists so that $\sum_{i \in R} v(\{i\}) < v(R)$ and $\sum_{i \in S} v(\{i\}) = v(S)$ for all other coalitions S of N.

Proof: If at least a $j \in R$ exists, so that $v(\{j\}) \cdot d_j^R < v(R)$, then at least one strategy x also exists, so that $\sum_{i \in R} x_i^R d_i^R = v(R)$.

This strategy is a Nash equilibrium, inasmuch as no player increases his/her payoff by abandoning it; moreover, bearing in mind what was said in Sect. 2.3, it leads to a

Pareto-optimal payoff. If, on the other hand, no j exists with such a property, then there is a sole TG-solution that consists in $p_i = v(\{i\})$. QED

Remark 3. As may be seen in the proof, the TG-solution for games of this type is made up of a subset of the core.

Proposition 4. The TG-solution is not empty for all 2-person games.

Proof: Depending on whether the game is subadditive, inessential, or superadditive, this can be proved by applying Propositions 1, 2, or 3, respectively. QED

Remark 4. The relationships between a TG-solution and the core have been covered within the preceding remarks.

Proposition 5. The TG-solution is not empty for all 3-person simple games.

Proof: We shall prove the proposition only for those cases not already covered by the preceding propositions.

Let i, j, k be the labels of the three players. We shall restrict ourselves to giving the solutions, since Nash equilibria and Pareto-optimality are self-evident.

- Case $v(N) = 0$

    If $v(\{i\}) = v(\{j\}) = 0$, $v(\{k\}) = 1$, $v(\{i, j\}) = 1$ and $v(\{i, k\}) = v(\{j, k\}) = 0$, then the TG-solution is the set of payoffs so that $p_k = 1$ and $p_i + p_j = 1$.
    If $v(\{j\}) = 1$, $v(\{k\}) = v(\{i\}) = 0$, $v(\{i, j\}) = v(\{i, k\}) = 1$ and $v(\{k, j\}) = 0$, then the TG-solution is the set of payoffs so that $p_j = 1$ and $p_i + p_k = 1$.
    If $v(\{i\}) = 1$, $v(\{j\}) = v(\{k\}) = 0$, $v(\{i, j\}) = v(\{i, k\}) = v(\{j, k\}) = 1$, then the TG-solution is the set of payoffs so that $p_i = 1$ and $p_j + p_k = 1$.

- Case $v(N) \le 1$

    If $v(\{i\}) = v(\{j\}) = v(\{k\}) = 0$, $v(\{i, j\}) = 1$, $v(\{i, k\}) = v(\{j, k\}) = 0$, then the TG-solution is the set of payoffs so that $p_k = 0$ and $p_i + p_j = 1$.
    If $v(\{i\}) = 1$, $v(\{j\}) = v(\{k\}) = 0$, $v(\{i, j\}) = v(\{i, k\}) = 0$, $v(\{j, k\}) = 1$, then the TG-solution is the set of payoffs so that $p_i = 0$ and $p_j + p_k = 1$.
    If $v(\{i\}) = v(\{k\}) = 0$, $v(\{j\}) = 1$, $v(\{i, j\}) = v(\{i, k\}) = 1$, $v(\{j, k\}) = 0$, then the TG-solution is the set of payoffs so that $p_j = 1$ and $p_i + p_k = 1$.
    For all other cases, the TG-solution corresponds to the set of feasible payoffs so that $p_i + p_j + p_k = 1$. QED

Remark 5. From this proof we can deduce that the TG-solution for every 3-person simple game having a not-empty core belongs to the core.

Definition. According to [2], we define *incore* of every game v the set of imputations p so that:

$$\sum_{i=1}^{n} p_i = v(N)$$

and

$$\sum_{i \in S} p_i > v(S) \text{ for all other } S \subset N.$$

Observe that the only limitations that affinities a_i^S must comply, are those quoted in Sect. 2.1. Of course, additional restrictions may be added during the application stage, but, in theory, the model allows any freedom to those indices. Namely, it allows that the affinity of each player with respect to a coalition be equal, or less, or greater, than its affinity with any other coalition. Therefore, the model can consider even cases where every player has the highest affinity towards the global coalition, as in the following:

Proposition 6. The TG-solution is not empty for every superadditive game v so that:

- the incore of v is not empty;
- $a_i^N = 1$.
- $a_i^S < 1$ for all coalitions S different to N and $\{i\}$.

Proof: As v has not empty incore, then a strategy x exists so that:

$$\sum_{i \in N} x_i = v(N)$$

while, for all S other than N,

$$\sum_{i \in S} x_i > v(S).$$

Hence:

$$\sum_{i \in N} x_i^N = \sum_{i \in N} x_i = v(N)$$

and

$$\sum_{i \in S} x_i^S \geq \sum_{i \in S} x_i > v(S).$$

Therefore, the only winning coalition is N and the payments are $p_i = x_i$ for all i. Such strategy is Pareto-optimal, since if player i asks for more, he receives zero. QED

Remark 6. From the above proof we can deduce that the TG-solution of all games respecting the conditions of Proposition 3 is a proper subset of their core.

6 Conclusions and Open Problems

The model proposed in Gambarelli [2] had returned strategies to the hands of players, even in games in characteristic function form. More precisely, in [1], von Neumann and Morgenstern introduced a method for transforming games from characteristic function form into normal form. This method is limited to constant-sum games. In [2], Gambarelli proposed a generalization of the von Neumann and Morgenstern method to variable-sum games.

In the current paper, the concept behind the solutions constructed in Gambarelli [2] has been extended to the case of differing affinities among players and coalitions, so as to bring the model closer to real-life situations. The non-emptiness of the TG-solution has been proved for several classes of games. Precisely, the TG-solution is not empty for the following class of games: subadditive, inessential, 2-person games, 3-person simple games, and also for all superadditive games satisfying particular condition (see Sect. 5). Finally, an algorithm for the automatic generation of the solution is supplied.

What remains, is to extend the study of the existence of the solution for other classes of games, to complete the connections with the core, and to test the validity of the solution in real life. More particularly, there might be interesting applications in simple games, in which various degrees of affinity are of greater importance: see, for instance, Bertini, Freixas, Gambarelli, and Stach [4], Gambarelli [5], and Gambarelli and Owen [6].

Acknowledgements. This paper is sponsored by MIUR, by research grants from the University of Bergamo, by the Group GNAMPA of INDAM and the statutory funds (no. 11/11.200.322) of the AGH University of Science and Technology. The authors thank Angelo Uristani for his useful suggestions.

Finally, the authors would like to thank the anonymous reviewers for their careful reading of our manuscript and their many insightful comments and suggestions.

Appendix

In this Appendix we reference a software program, written in Python 2.7.6, for the generation of the TG-solution. Information concerning computation times, and the number of found vectors for the solution, are given further below.

```
from itertools import combinations, product
from fractions import Fraction

def strif(string):          #to integer or float
  f = float(string)
  i = int(f)
  return i if f == i else f

#input
n = 0    #number of players
while n < 2:
  try: n = int(raw_input('How many players are there in
this game (minimum 2)? '))
  except ValueError: print 'ERROR! The number of players
must be expressed by an integer bigger than 1!'

players = range(1,n+1)
value = {}     #building characteristic function
players_comb = [comb for length in players for comb in
combinations(players,length)]
for coal in players_comb:
  while coal not in value:
    try:
      value[coal] = Fraction(raw_input("v(%s): what's the
value of this coalition? " % ','.join(map(str,coal))))
    except ValueError:
      print "ERROR! The value of the coalition must be
expressed in numbers!"

a = {}       #affinity
twoPlayersCoal = players_comb[n:n+(n*(n-1))/2]
```

```python
moreThanTwoPlayersCoal = players_comb[n+(n*(n-1))/2:]
for player in players:
  a[player] = {(player,): 1}
  print "What's the level of affinity (min 0 - max 1) of
player %d towards the following coalitions?" % player
  for coal in twoPlayersCoal:
    if player in coal:
      while coal not in a[player]:
        try:
          affinity = Fraction(raw_input("(%s): " %
',' .join(map(str,coal))))
          if affinity < 0 or affinity > 1: raise ValueEr-
ror
          else: a[player][coal] = affinity
        except ValueError:
          print "ERROR! The affinity must be a decimal
value between 0 and 1!"
  for coal in moreThanTwoPlayersCoal:
    if player in coal:
      a[player][coal] = min(a[player][x] for x in
a[player] if all(member in coal for member in x))

while True:   #level of precision
  try:
    precision = Fraction(raw_input('What\'s the interval
of analysis desired (it must be a divisor of each coali-
tion value)? '))
    if all(value % precision == 0 for value in value.val-
ues()): break
    else: raise ValueError
  except ValueError:
    print 'ERROR! You must input a positive number divi-
sor of each coalition value!'
#identification of the strategies
strategies = {}
for player in players:
  coalwithplayer = [coal for coal in players_comb if
player in coal]
  limit = max(value[coal] for coal in coalwithplayer)
  while not any(value[coal] >= limit*Frac-
tion(1,a[player][coal]) + sum(value[(member,)]*Frac-
tion(1,a[member][coal]) for member in coal if member !=
player) for coal in coalwithplayer): limit -= precision
```

```
  strategies[player] = [strif(value[(player,)] + x*preci-
sion) for x in xrange(int((limit-value[(player,)])/preci-
sion)+1)]

#identification of the pwc
PWC = {}
allocations = list(product(*[strategies[player] for
player in players]))
max_width = max(len(str(alloc)) for alloc in alloca-
tions)+15
for alloc in allocations:
  PWC[alloc] = [coal for coal in players_comb if
value[coal] >= sum(alloc[player-1]*Frac-
tion(1,a[player][coal]) for player in coal)]

#identification of the pwl
H = [coalset for l in players for coalset in list(combi-
nations(players_comb,l)) if sum(map(len,coalset)) == n
and set(player for coal in coalset for player in coal) ==
set(players)]
PWL = {}
for alloc in allocations:
  layouts = set(tuple(coal for coal in coalset if coal in
PWC[alloc]) for coalset in H)
  PWL[alloc] = [list(layout) for layout in layouts if
sum(1 for otherlayout in layouts if set(layout) <=
set(otherlayout)) <= 1]

#computation of the payments
P = {}
for alloc in allocations:
  probs = [len(PWL[alloc])]+[sum(1 for layout in
PWL[alloc] if any(player in coal for coal in layout)) for
player in players]
  if probs[0] != 0: P[alloc] = [Fraction(Frac-
tion(probs[player],probs[0])*alloc[player-1]) for player
in players]
  else: P[alloc] = [0]*n

#identification of the pareto-efficient payments
POP = {}
for alloc in allocations:
  for otheralloc in P:
    if any(P[otheralloc][player] > P[alloc][player] for
player in xrange(n)):
```

```
    if not any(P[otheralloc][player] < P[alloc][player]
for player in xrange(n)): break
  else: POP[alloc] = P[alloc]

#identification of the NE
print 'This is the TG-solution:'
for alloc in sorted(POP):
  if all(P[otheralloc][player] <= POP[alloc][player] for
otheralloc in P.iterkeys() for player in xrange(n) if
all(otheralloc[pl] == alloc[pl] for pl in xrange(n) if pl
!= player)):
    if all(-0.1*precision <= sum(alloc[player-1]*Frac-
tion(1,a[player][pwc]) for player in pwc) - value[pwc] <=
0.1*precision for pwc in PWC[alloc]):
      print str(alloc).ljust(max_width), map(lambda x:
round(strif(x),2),P[alloc])
```

On the Computation Time

We carried out the calculation of the solution with different values of n and δ. The considered games were the following:

G2: $n = 2$, $v(\{1\}) = v(\{2\}) = 0$, $v(\{1,2\}) = 1$;

G3: $n = 3$, $v(S) =$ the same of G2, and $v(\{3\}) = 1$, $v(\{1,3\}) = 1$, $v(\{2,3\}) = 2$, $v(\{1,2,3\}) = 3$ (it is the example of Sect. 4);

G4: $n = 4$, $v(S) =$ the same of G3, and $v(\{4\}) = 0$, $v(\{1,4\}) = v(\{2,4\}) = 0$, $v(\{3,4\}) = 1$, $v(\{1,2,4\}) = v(\{1,3,4\}) = v(\{2,3,4\}) = 1$, $v(\{1,2,3,4\}) = 3$

We used the same affinities reported in Table 1; for all other cases we set $a_i^S = 1$.

The computation has been undertaken using an ACER Aspire 3935 computer, with 4 GB DDR3 Memory, Intel CoreTM 2 Duo processor P7450 (2,13 GHz, 1066 MHz FSB).

OS: Ubuntu 14.04.

The numbers of found vectors for the TG-solution and the computation times are shown in Tables 5 and 6.

Table 5. Number of found vectors for the TG-solution.

G	δ		
	1/3	1/5	1/10
G2	1	1	1
G3	0	0	5
G4	1	2	82

Table 6. Computation times (in seconds).

G	δ		
	1/3	1/5	1/10
$G2$	0.00	0.00	0.02
$G3$	0.02	0.06	0.47
$G4$	0.13	1.14	64.87

References

1. Von Neumann, J., Morgenstern, O.: Theory of Games and Economic Behavior, 3rd edn. 1953. Princeton University Press, Princeton (1944)
2. Gambarelli, G.: Transforming games from characteristic into normal form. In: Vannucci, S., Ursini, A. (eds.) International Game Theory Review. Special Issue devoted to Logic, Game Theory and Social Choice, vol. 9(1), pp. 87–104 (2007)
3. Aumann, R.J.: A survey of cooperative games without side payments. In: Shubik, M. (ed.) Essays in Mathematical Economics, pp. 3–27. Princeton University Press, Princeton (1967)
4. Bertini, C., Freixas, J., Gambarelli, G., Stach, I.: Some open problems in simple games. In: Fragnelli, V., Gambarelli, G. (eds.) Open Problems in the Theory of Cooperative Games. Special Issue of International Game Theory Review, vol. 15(2), pp. 1340005-1–1340005-18 (2013)
5. Gambarelli, G.: Common behavior of power indices. Int. J. Game Theory 12(4), 237–244 (1983)
6. Gambarelli, G., Owen, G.: Indirect control of corporations. Int. J. Game Theory 23(4), 287–302 (1994)

Comparing Game-Theoretic and Maximum Likelihood Approaches for Network Partitioning

Vladimir V. Mazalov[1,2(✉)]

[1] Institute of Applied Mathematical Research,
Karelian Research Center, Russian Academy of Sciences,
11, Pushkinskaya st., Petrozavodsk 185910, Russia
vmazalov@krc.karelia.ru
[2] School of Mathematcis and Statistics and Institute of Applied Mathematcis,
Qingdao University, Qingdao 266071, People's Republic of China

Abstract. The purpose of this article is to show the relationship between the game-theoretic approach and the maximum likelihood method in the problem of community detection in networks. We formulate a cooperative game related with network structure where the nodes are players in a hedonic game and then we find the stable partition of the network into coalitions. This approach corresponds to the problem of maximizing a potential function and allows to detect clusters with various resolution. We propose here the maximum likelihood method for the tuning of the resolution parameter in the hedonic game. We illustrate this approach by some numerical examples.

Keywords: Network partitioning · Community detection
Cooperative games · Hedonic games · Tuning of the parameter
Maximum likelihood method

1 Introduction

Many complex systems can be represented as networks which contain more than tens or hundreds of thousands of nodes. In these networks the users are the nodes and their mutual interactions are links of a system. Complex systems are usually structured as a set of subsystems which have their own function. Inside the component the nodes have a high density of internal links, whereas links between components have a significantly lower density. This feature can be used for partitioning of the network into components.

Network partitioning or community detection in networks is a very important topic which attracted the effort of many researchers. There are some popular approaches for network partitioning.

The first very large class is based on spectral elements of the Laplacian matrix (see the survey [19] and references therein, and [13]). The second class of methods

© Springer-Verlag GmbH Germany, part of Springer Nature 2018
N. T. Nguyen et al. (Eds.): TCCI XXXI, LNCS 11290, pp. 37–46, 2018.
https://doi.org/10.1007/978-3-662-58464-4_4

is based on the use of random walks (see e.g., [1,2,8,14,16]). The third class of approaches to network partitioning is based on methods from statistical physics [4,17,18]. The fourth class, which is based on the concept of modularity and its various generalizations [5,11,15,20]. The overview of the community detection methods you can see in the survey [10].

In the present work, we explore the approach based on hedonic games [6], which are games explaining the mechanism behind the formation of coalitions. This approach allows to detect components of the network with varying resolution. The parameter of resolution is very important factor for the partitioning. In [3] it was shown that the hedonic game approach is especially well suited to adjust the level of resolution as the limiting cases are given by the grand coalition and maximal clique decomposition, two very natural extreme cases of network partitioning. Here we propose the maximum likelihood method for the tuning of the resolution parameter in the hedonic game.

The paper is structured as follows: in the Sect. 2 we formally define network partitioning as a cooperative game. Then, we present our approach based on the hedonic games. In Sect. 3 we define the maximum likelihood method to find the Nash stable partition. Also we provide illustrative examples which explain the essence of the methods. In Sect. 4 we propose a method for tuning of the resolution parameter in the hedonic game. Finally, Sect. 5 concludes the paper with the algorithm for network partitioning.

2   Game-Theoretic Approach for Clustering

Network clustering problem can be considered as a community detection problem. In this approach we consider the nodes of the network as players in some game in which the payoffs of the players are determined via structure of the network. Following this way we can determine the preferences of the players to be included in different coalitions. We are interested in stable partitions of the network into the communities of the players.

Originally, we have a network $G = (N, E)$ where $N = \{1, 2, ..., n\}$ is set of nodes (players) and E is a set of edges (links). Denote $E(i, j)$ the weight of the link $e = (i, j)$. If $(i, j) \notin E$ then $E(i, j) = 0$. For unweighted network $E(i, j) = 1$ if $(i, j) \in E$.

In paper [3] it was presented game-theoretic approach for the partitioning of a society into coalitions based on the definition of hedonic games [6]. The framework of hedonic games is specifying the preference function.

Assume that the set of players $N = \{1, \ldots, n\}$ is divided into K coalitions: $\Pi = \{S_1, \ldots, S_K\}$. Let $S_\Pi(i)$ denote the coalition $S_k \in \Pi$ such that $i \in S_k$. A player i preferences are represented by a complete, reflexive and transitive binary relation $\succeq_i$ over the set $\{S \subset N : i \in S\}$. The preferences are additively separable [6] if there exists a value function $v_i : N \to \mathbb{R}$ such that $v_i(i) = 0$ and

$$S_1 \succeq_i S_2 \Leftrightarrow \sum_{j \in S_1} v_i(j) \geq \sum_{j \in S_2} v_i(j).$$

The preferences $\{v_i, i \in N\}$ are symmetric, if $v_i(j) = v_j(i) = v_{ij} = v_{ji}$ for all $i, j \in N$. The symmetry property defines a very important class of hedonic games.

We call the network partition Π is *Nash stable*, if $S_\Pi(i) \succeq_i S_k \cup \{i\}$ for all $i \in N, S_k \in \Pi \cup \{\emptyset\}$. In the Nash-stable partition, there is no player who wants to leave her coalition. A potential of a coalition partition $\Pi = \{S_1, \ldots, S_K\}$ (see [6]) is

$$P(\Pi) = \sum_{k=1}^{K} P(S_k) = \sum_{k=1}^{K} \sum_{i,j \in S_k} v_{ij}. \tag{1}$$

Our method for detecting a stable community structure is based on the following better response type dynamics.

Start with any partition of the network $N = \{S_1, \ldots, S_K\}$. Choose any player i and any coalition S_k different from $S_\Pi(i)$. If $S_k \cup \{i\} \succeq_i S_\Pi(i)$, assign node i to the coalition S_k; otherwise, keep the partition unchanged and choose another pair of node-coalition, etc.

Since the game has the potential (1), the above algorithm is guaranteed to converge in a finite number of steps [12].

One natural way to define a symmetric value function v with a parameter $\alpha \in [0, 1]$ is as follows:

$$v_{ij} = E(i, j) - \alpha. \tag{2}$$

For any subgraph $(S, E|S)$, $S \subseteq N$, denote $n(S)$ as the number of nodes in S, and $m(S)$ as the number of edges in S. Then, for the value function (2), the potential (1) takes the form

$$P(\Pi) = \sum_{k=1}^{K} \left(m(S_k) - \frac{n(S_k)(n(S_k) - 1)\alpha}{2} \right). \tag{3}$$

Notice that still exists the uncertainty in the selection of the parameter α.

Example 1. Consider the network of eighteen nodes presented in Fig. 1. The network consists of two cliques: $G_1 = \{A, B, C, D, E\}$ and $G_2 = \{J, K, L, M, N\}$. In the network there are ten edges between G_1 and G_2, i.e. (u, w), such that $u \in G_1$ and $w \in G_2$. Also, there are four nodes connected with the nodes from G_1, and four nodes connected with G_2.

A natural partition of this network into two clusters is the partition $\Pi = S_L \cup S_R = \{A, B, ..., H, I\} \cup \{J, K, ..., R, S\}$: one cluster is at the right side of the picture and another one is at the left side.

From formula (3) we find the potential

$$P(S_L \cup S_R) = 28 - 72\alpha.$$

Assume that the node J joins to the left coalition. Then

$$P(\{S_L \setminus J\} \cup \{S_R \cup J\}) = 26 - 73\alpha.$$

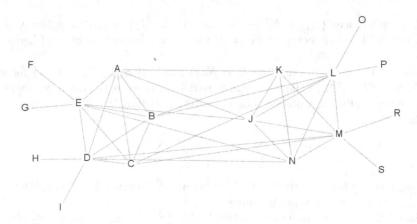

Fig. 1. Network of eighteen nodes

We see that for any $\alpha \geq 0$ the partition $S_L \cup S_R$ is more preferable for player J than the partition $\{S_L \setminus J\} \cup \{S_R \cup J\}$. So, for player J there is no incentive to join to the left coalition.

We can show that $P(S_L \cup S_R) > P(\Pi')$, where Π' is any partition which can be obtained from $S_L \cup S_R$ by moving of one node from one cluster to another cluster. That is true for any value of α. It means that $S_L \cup S_R$ is Nash stable partition in the hedonic game with potential (3).

Notice that there are also another stable partitions. For example we can show that the partition

$$\Pi^* = \{A,B,C,D,E,J,K,L,M,N\}\cup\{F\}\cup\{G\}\cup\{H\}\cup\{I\}\cup\{O\}\cup\{P\}\cup\{R\}\cup\{S\}$$

is Nash stable. By formula (3) we find

$$P(\Pi^*) = 30 - 45\alpha.$$

Comparing with $P(S_L \cup S_R) = 28 - 72\alpha$ we find that $P(\Pi^*) > P(S_L \cup S_R)$ for all α.

Now consider the partition

$$\Pi^{**} = \{A,B,C,D,E\}\cup\{J,K,L,M,N\}\cup\{F\}\cup\{G\}\cup\{H\}\cup\{I\}\cup\{O\}\cup\{P\}\cup\{R\}\cup\{S\}.$$

That is also Nash stable partition. By formula (3) we find

$$P(\Pi^{**}) = 20 - 20\alpha.$$

Comparing the partitions Π^* and Π^{**} we conclude that the partition Π^* is optimal (in a sense of potential function) for $\alpha \leq 2/5$, and the partition Π^{**} is optimal for $\alpha > 2/5$.

We see that selection of the parameter α is significantly affected for the network partitioning. In next section we propose the idea how to select the value of the parameter α.

3   Maximum Likelihood Method for Clustering

Here we use the probabilistic approach for network partitioning. The idea is closed to [7]. Suppose that the network is generated with some inherent randomness. Introduce two parameters: p_{in} is the probability of any two given nodes inside a community interacting and p_{out} is the probability of two given nodes from different communities interacting. The "randomness" is determined by these probabilities p_{in} and p_{out}. Maximizing the most probable partition in all possible community structures we can find the most likely community structure which corresponds the real data.

Consider a network of $N = \{1, 2, ..., n\}$ nodes (players). Let $M = m(E)$ is the number of all edges in network. Denote $\Pi(N)$ the set of all partitions of N.

Assume that there is a true coalition partition of the network $\Pi = \{S_1, ..., S_K\}$. Let $n_k = n(S_k)$ and $m_k = m(S_k)$ denote the number of the nodes and edges in the coalition $S_k, k = 1, ..., K$, respectively. So, $n = \sum_{k=1}^{K} n_k$ and $\sum_{k=1}^{K} m_k \leq M$.

Consider a coalition $S_k \in \Pi$. The probability of implementation exactly m_k connections among n_k players in the coalition S_k is equal to

$$p_{in}^{m_k} (1 - p_{in})^{\frac{n_k(n_k-1)}{2} - m_k}.$$

Each player i from coalition S_k can have the in general $M - m_k$ connections with the players from other coalitions but in fact he has $\sum_{j \notin S_k} E(i, j)$ connections with the players from other coalitions.

The probability of implementation of the network with given structure is equal to

$$L_\Pi = \prod_{k=1}^{K} p_{in}^{m_k} (1 - p_{in})^{\frac{n_k(n_k-1)}{2} - m_k} \prod_{i \in S_k} p_{out}^{\frac{1}{2} \sum_{j \notin S_k} E(i,j)} (1 - p_{out})^{\frac{1}{2}(M - m_k - \sum_{j \notin S_k} E(i,j))}.$$

(4)

By taking logs of likelihood function L_Π from (4) and simplifying we obtain

$$l_\Pi = \log L_\Pi = \sum_{k=1}^{K} m_k \log p_{in} + \sum_{k=1}^{K} \left(\frac{n_k(n_k - 1)}{2} - m_k \right) \log(1 - p_{in}) +$$

$$(M - \sum_{k=1}^{K} m_k) \log p_{out} + \left(\frac{1}{2} \sum_{k=1}^{K} n_k(n - n_k) - (M - \sum_{k=1}^{K} m_k) \right) \log(1 - p_{out}). (5)$$

The partition Π^* for which the function l_Π achieves its maximum we call the optimal partition. Notice that exists an uncertainty in the selection of the probabilities p_{in} and p_{out}. The function $l_\Pi = l_\Pi(p_{in}, p_{out})$ is a function of the arguments p_{in}, p_{out}. Maximizing l_Π in p_{in}, p_{out} we can use it in numerical simulations.

Proposition 1. For fixed partition Π the function $l_\Pi(p_{in}, p_{out})$ attains the maximum when

$$p_{in} = \frac{2\sum_{k=1}^{K} m_k}{\sum_{k=1}^{K} n_k^2 - n}, \qquad p_{out} = \frac{2(M - \sum_{k=1}^{K} m_k)}{n^2 - \sum_{k=1}^{K} n_k^2}. \tag{6}$$

Substituting (6) into (5) we obtain the expression which depends only of the structure of the network.

Example 2. Consider the simple network of six nodes presented in Fig. 2.

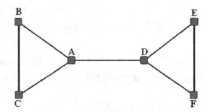

Fig. 2. Network of six nodes

Let us calculate the value of l_Π for different partitions. For partition $\Pi = \{B, C\} \cup \{A, D\} \cup \{E, F\}$ we obtain from (5)

$$l_\Pi = 3\log p_{in} + 4\log p_{out} + 8\log(1 - p_{out}).$$

Maximum of this function attains for $p_{in} = 1$ and $p_{out} = 1/3$. The maximal value is -7.638.

For partition $\Pi = \{A, B, C, D\} \cup \{E, F\}$ it yields

$$l_\Pi = 5\log p_{in} + 2\log(1 - p_{in}) + 2\log p_{out} + 6\log(1 - p_{out}).$$

Maximum of this function attains for $p_{in} = 5/7$ and $p_{out} = 1/4$. The maximal value is -8.686.

For partition $\Pi = \{A, B, C\} \cup \{D, E, F\}$ we obtain

$$l_\Pi = 6\log p_{in} + \log p_{out} + 8\log(1 - p_{out}).$$

Maximum of this function attains for $p_{in} = 1$ and $p_{out} = 1/9$. The maximal value is -3.139.

We see that the partition $\Pi = \{A, B, C\} \cup \{D, E, F\}$ gives the most probable community structure for this network.

4   Connection of Maximum Likelihood and Game-Theoretic Methods

First, present the function (5) as

$$l_\Pi = \sum_{k=1}^{K} m_k \log \frac{p_{in}(1-p_{out})}{p_{out}(1-p_{in})} - \frac{1}{2} \sum_{k=1}^{K} n_k^2 \log \frac{1-p_{out}}{1-p_{in}} + R, \qquad (7)$$

where the term

$$R = -\frac{n}{2} \log(1-p_{in}) + M \log p_{out} + (\frac{1}{2}n^2 - M) \log(1-p_{out})$$

depends on the characteristics of the network. It is convenient to present (7) as

$$l_\Pi = \log \frac{p_{in}(1-p_{out})}{p_{out}(1-p_{in})} \left(\sum_{k=1}^{K} m_k - \frac{1}{2} \sum_{k=1}^{K} n_k^2 \frac{\log \frac{1-p_{out}}{1-p_{in}}}{\log \frac{p_{in}(1-p_{out})}{p_{out}(1-p_{in})}} \right) + R'. \qquad (8)$$

Now return back to the potential (3). Rewrite it in the form

$$P(\Pi) = \left(\sum_{k=1}^{K} m_k - \frac{1}{2} \sum_{k=1}^{K} n_k^2 \alpha \right) + \frac{1}{2} n\alpha. \qquad (9)$$

Comparing (8) and (9) we see that the role of α in (9) belongs to the factor in the brackets if formula (8).

Proposition 2. The maximum likelihood and game-theoretic methods are equivalent if the resolution parameters are connected with formula

$$\alpha = \frac{\log \frac{1-p_{out}}{1-p_{in}}}{\log \frac{p_{in}(1-p_{out})}{p_{out}(1-p_{in})}} = \frac{\log \frac{1-p_{out}}{1-p_{in}}}{\log \frac{p_{in}}{p_{out}} + \log \frac{1-p_{out}}{1-p_{in}}}. \qquad (10)$$

We can use this fact for tuning of the parameter α in the algorithm based on the hedonic games theory.

5   Algorithm of Nash Stable Partitioning

Describe the algorithm which can find the Nash stable partition on the base of hedonic games with objective function (9). Stable partition is the partition for which the objective function reaches a local maximum.

We start from a partition Π_0. In each iteration we consider the current partition Π and partition Π' which is obtained by moving of one player $i \in S_{\Pi(i)}$ into coalition $S_k \in \Pi$. In the coalition S_k the player i has got $d_i(S_k \cup i)$ connections with weight $1 - \alpha$ minus $m(S_k) + 1 - d_i(S_k \cup i)$ connections with weight α. At the same time player i has lost in previous coalition $d_i(S_{\Pi(i)})$ connections with weight $1 - \alpha$ minus $m(S_{\Pi(i)}) - d_i(S_{\Pi(i)})$ connections with weight α. The parameter α we estimate using the formula (10).

Thus, the variation of potential function is

$$P(\Pi') - P(\Pi) = d_i(S_k \cup i) - d_i(S_{\Pi(i)}) + \alpha(m(S_{\Pi(i)}) - m(S_k) - 1).$$

In the case when the player i becomes a single the variation of potential function is

$$P(\Pi') - P(\Pi) = -d_i(S_{\Pi(i)}) + \alpha(m(S_{\Pi(i)}) - 1).$$

Among all possible partitions Π' we select a partition such that $P(\Pi') - P(\Pi) \to max$. If $P(\Pi') - P(\Pi) > 0$, then the current partition is changed for Π', otherwise the partition Π is Nash stable partition.

The Algorithm

The community detection algorithm is described in Algorithm 1. Initially all nodes are placed in K communities, K is fixed. We need to define the parameters:

$N = \{1, 2, ..., n\}$, the set of players

M, number of edges

K, number of communities

$n_k, m_k, \quad k = 1, ..., K$, number of nodes and edges in each coalition

T, number of stages. In each stage we recalculate p_{in}, p_{out} and α

L, number of iterations in each stage.

Algorithm 1. Community Detection

Input : $\{L\}_{t=1}^T$
Output: Π
1 $\Pi \leftarrow$ arbitrary initial partition Π_0
2 **for** t **from** 1 **to** T **do**
3    $\Pi \leftarrow$ main (Π, L_t)
4 **end**
5 **return** Π

The Algorithm 2 is responsible for selection of the records for the objective function $P(\Pi)$. It uses the Algorithm 3 which generates a new partition Π'. In each iteration a player $i \in N$ can move with probability p from the current coalition to any coalition $k \in \{1, ..., K\}$.

Algorithm 2. main

Input : $\{\Pi, L\}$
Output: Π'
1 $P^* \leftarrow P(\Pi)$
2 **for** l **from** 1 **to** L **do**
3    $\Pi' \leftarrow \text{gen}(\Pi, l)$
4   **if** $P(\Pi) < P(\Pi')$ **then**
5       $\Pi \leftarrow \Pi'$
6       $P^* \leftarrow \max(P^*, P(\Pi))$
7 **end**
8 **return** Π

Algorithm 3. gen

Input : $\{\Pi, l\}$
Output: Π'
1 $p \leftarrow (l \mod (n+1)/n)$
2 **for** $i \in N$ **do**
3   **if** random $\xi < p$ **then**
4       $\Pi(i) \leftarrow$ one random from $\{1, 2, ..., K\}$
5   **end**
6 **end**
7 **return** Π

---

6 Conclusion

We have presented the approach based on the hedonic game theory and maximum likelihood method for network partitioning. We show there is a connection between these approaches and use it for the tuning of the parameters of the methods. Some numerical examples are presented.

The efficiency of the game-theoretic approach for large networks was demonstrated in paper [9]. Finding Nash equilibrium is equivalent to finding a maximum of the game's potential. It was shown that the maximization problem can be numerically solved by the algorithm which was presented here. The efficiency of the method was demonstrated on the fragment of the Russian academic Web which consists of the official sites of the Siberian and Far East branches of the Russian Academy of Science. This network contains 140 scientific sites connected by 2315 links. Numerical analysis gives the partition of the academic Web into two main components corresponding to Siberian and Far East divisions of the Russian Academy of Science.

The selection of the resolution parameter in the experiments was intuitive. Here we derive the explicit formula how this parameter can be determined.

Acknowledgements. This research is supported by the Russian Fund for Basic Research (projects 16-51-55006, 16-01-00183) and Shandong Province "Double-Hundred Talent Plan (No. WST2017009)".

References

1. Avrachenkov, K., Dobrynin, V., Nemirovsky, D., Pham, S.K., Smirnova, E.: Pagerank based clustering of hypertext document collections. Proc. ACM SIGIR **2008**, 873–874 (2008)
2. Avrachenkov, K., El Chamie, M., Neglia, G.: Graph clustering based on mixing time of random walks. Proc. IEEE ICC **2014**, 4089–4094 (2014)
3. Avrachenkov, K.E., Kondratev, A.Y., Mazalov, V.V.: Cooperative game theory approaches for network partitioning. In: Cao, Y., Chen, J. (eds.) COCOON 2017. LNCS, vol. 10392, pp. 591–602. Springer, Cham (2017). https://doi.org/10.1007/978-3-319-62389-4_49
4. Blatt, M., Wiseman, S., Domany, E.: Clustering data through an analogy to the Potts model. In: Proceedings of NIPS 1996, pp. 416–422 (1996)
5. Blondel, V.D., Guillaume, J.L., Lambiotte, R., Lefebvre, E.: Fast unfolding of communities in large networks. J. Stat. Mech. Theory Exp. **10**, 10008 (2008)
6. Bogomolnaia, A., Jackson, M.O.: The stability of hedonic coalition structures. Games Econ. Behav. **38**(2), 201–230 (2002)
7. Copic, J., Jackson, M., Kirman, A.: Identifying community structures from network data via maximum likelihood methods. B.E. J. Theor. Econ. **9**(1) (2009). 1935-1704
8. Dongen, S.: Performance criteria for graph clustering and Markov cluster experiments, CWI Technical report (2000)
9. Ermolin, N.A., Mazalov, V.V., Pechnikov, A.A.: Game-theoretic methods for finding communities in academic Web. In: SPIIRAS Proceedings, Issue 6(55), pp. 237–254 (2017)
10. Fortunato, S.: Community detection in graphs. Phys. Rep. **486**(3), 75–174 (2010)
11. Girvan, M., Newman, M.E.J.: Community structure in social and biological networks. Proc. Nat. Acad. Sci. USA **99**(12), 7821–7826 (2002)
12. Mazalov, V.: Mathematical Game Theory and Applications. Wiley, Hoboken (2014)
13. Mazalov, V.V., Tsynguev, B.T.: Kirchhoff centrality measure for collaboration network. In: Nguyen, H.T.T., Snasel, V. (eds.) CSoNet 2016. LNCS, vol. 9795, pp. 147–157. Springer, Cham (2016). https://doi.org/10.1007/978-3-319-42345-6_13
14. Meila, M., Shi, J.: A random walks view of spectral segmentation. In: Proceedings of AISTATS (2001)
15. Newman, M.E.J.: Modularity and community structure in networks. Soc. Netw. **103**(23), 8577–8582 (2006)
16. Pons, P., Latapy, M.: Computing communities in large networks using random walks. J. Graph Algo. Appl. **10**(2), 191–218 (2006)
17. Raghavan, U.N., Albert, R., Kumara, S.: Near linear time algorithm to detect community structures in large-scale networks. Phys. Rev. E. **76**(3), 036106 (2007)
18. Reichardt, J., Bornholdt, S.: Statistical mechanics of community detection. Phys. Rev. E. **74**(1), 016110 (2006)
19. von Luxburg, U.: A tutorial on spectral clustering. Stat. Comput. **17**(4), 395–416 (2007)
20. Waltman, L., van Eck, N.J., Noyons, E.C.: A unified approach to mapping and clustering of bibliometric networks. J. Inform. **4**(4), 629–635 (2010)

Comparing Results of Voting by Statistical Rank Tests

Krzysztof Przybyszewski[1] and Honorata Sosnowska[2]($\boxtimes$)

[1] Leon Kozminski Academy, Warsaw, Poland
[2] Warsaw School of Economics, Warsaw, Poland
honorata@sgh.waw.pl

Abstract. In the paper the results of voting by different voting methods are presented and analyzed. Approval voting, disapproval voting, categorization method and classical majority voting are compared using results of a presidential poll conducted over a representative sample. Results differ depending on the voting methods used. However, from the statistical point of view, within the same valuation scope (positive, if voting is constructed so that the vote is "for" the candidate and negative, if the vote is "against") the order of the candidates remains similar across the methods.

Keywords: Votings · Experiments · Statistical rank tests

1 Introduction

The aim of this paper is to compare voting methods, especially these based on approval voting and find differences and similarities which can influence results. It is assumed that voters' preferences are the same in all votings. It is a well known fact, confirmed both mathematically and empirically ([1–3]) that different voting methods lead to different results. We analyze how results differ between some methods constructed on a similar basis in comparison with the classical majority method. In spite of the differences in results, there are some similarities when they are compared from a statistical point of view.

Four voting methods are considered as basic methods. Classical majority method (plurality method), where each voter can choose no more than one alternative and the alternative with the highest score wins. This is the most frequently used method. Approval voting method, where each voter can choose as many alternatives as he/she approves. The alternative with the highest score wins. This method was introduced by Brams and Fisburn ([4, 5]) and was used by the Security Council as an additional method, to narrow down the list of candidates for Secretary General ([6]). It is also often used by scientific societies. Disapproval voting method where each voter chooses alternatives which he/she disapproves. The alternative with the lowest score wins. Any electoral system that allows voter to express his/her disapproval may be called disapproval voting. The method used in this paper was introduced by Holubiec and Mercik ([2]). From the mathematical point of view, this method may be viewed as the reverse of approval voting. Categorization method, which was introduced as the

© Springer-Verlag GmbH Germany, part of Springer Nature 2018
N. T. Nguyen et al. (Eds.): TCCI XXXI, LNCS 11290, pp. 47–54, 2018.
https://doi.org/10.1007/978-3-662-58464-4_5

"combined approval and disapproval voting" (CAV) by Fensenthal in 1989 ([7]), where each voter assigns to every alternative a plus (+), a minus (−) or marks it as" not signed" (0). Then the balance of values is computed as the number of (+) minus the number of (−). This method was considered as dis&approval voting by Alcantud and Laruelle ([8]), and as evaluative voting by Hillinger ([9, 10]). The psychological studies on the differences in decision making processes under different voting rules are scarce. Most of the psychological literature focuses on the way the political attitudes, and subsequently preferences, are formed. However, there is some evidence for differences in mental processes evoked by different voting procedures. These differences are connected with cognitive effort in each method. From this perspective, a choice consists of a set or a sequence of mental operations and motivational processes, whose contents can be traced, or their parameters (i.e. the amount of information gathered before the choice is made, and time spent on the choice) may be measured. Voting methods differ in the task that a voter is to perform. The majority rule requires the choice of a single, "best" alternative. Both the approval and disapproval methods require the voter to divide the alternatives into two categories ("approved" and "other", or "disapproved" and "other"). The process required by the categorisation method is the most complex among the methods based on the approval voting procedure, as it requires the voter to sort the alternatives into "approved", "disapproved" and "other". Malawski et al. ([11]) tested the differences in the complexity of the cognitive processes between the decisions made under majority, approval and categorisation rules. The process tracing study employed the information matrix with cells with drop- down lists. These had to be opened by the decision-makers to reveal the data and this way the amount of data processed was measured. The results showed that the number of cells (the amount of data indicator) was highest in the case of the most complex categorisation method, and the majority and approval votings were not very different in this respect. In another study, the relationship was tested by Przybyszewski and Sosnowska ([12]) between the authoritarianism as a psychological measure of the complexity of thought (a tendency to choose less complex explanation of the situation, see ([13]) and the usage of approval voting. The results are consistent with the previous findings − people with less complex political views tend to use the opportunity to divide the alternatives into the categories offered by the approval voting (measured by the number of chosen alternatives in the approval voting method) significantly less than those with more complex (less authoritarian) minds.

The methods and scores based on the four presented above may be divided into 3 groups: with positive, negative and neutral valuation. A positive valuation characterizes methods or scores where voters choose alternatives which they approve. Classical majority method (plurality), approval voting, number of pluses and balance in categorization method are such methods. A negative valuation characterize methods or scores where voters choose alternatives which they disapprove. Disapproval voting and number of minuses in categorization method are such a method and score. Methods and scores where voters neither approve nor disapprove have a neutral valuation. The number of 0's in categorization method and the number of voters minus sum of approval and disapproval votes are such methods.

We conducted a voting experiment over a representative sample where voters voted in a presidential poll 4 times, using all 4 methods mentioned above. We compare the results of votings. There are some differences, especially the 3 first places, but the results are similar at certain aspects. We constructed a list of ordered alternatives for each method from the best to the worst based on one of the 4 basic methods mentioned above. Every two lists were compared using statistical rank tests, in this case Kendall's tau and Spearman rho ([14–16]). The lists within the same valuation scope (positive – vote is "for" a candidate, negative – it is "against" a candidate, neutral – voter does not have an opinion about a candidate) were positively correlated (tau greater than 50%, usually much greater). The lists from the reverse valuation scopes were weaker or not correlated. It shows that, in spite of some differences, within the same valuation scope, the lists (so the results of votings) are similar.

The paper is constructed as follows. Presidential poll is presented in Sect. 2. Preliminary information about the Kendall and Spearman rank test are provided in Sect. 3. Similarities of the results within the neutral valuation scope are presented in Sect. 4 and among positive and negative scope in Sect. 5. Section 6 contains conclusions.

2 Presidential Poll

The Polish presidential poll was conducted over a representative sample of 1053 individuals in March 2017 by ARIADNA poll agency. The fact that we analyzed a representative sample is important. Our experience shows that experiments on students may be useful as pilot studies, but may not reveal some properties because of homogeneity of the group ([13]).

The aim of the poll was not to predict the results of the presidential elections (planned for 2020 and individual preferences may change many times) but to analyze the approval voting method in comparison with other methods. Four methods, presented in Sect. 1, were investigated.

Voters voted 4 times. There were buffers between votings. The buffers were constructed on a sequence of problems connected with decision making under risk. Voters could not see their previous votings. Voters who participate in polls conducted by ARIADNA poll agency may win a small gadget as a prize for attendance.

There were 17 candidates. In what follows, they are marked by initials as we do not want to analyze the results from the political point of view. Table 1 presents results where candidates are ordered alphabetically. "Maj" denotes the classical majority method, "App." – approval voting, "Disapp." – disapproval voting method and "Cat." – categorization method divided into, +, –, 0 and balance.

First glance at Table 1 ("other" and "do not know" denote the number of voters who marked these answers in the questionnaire) shows that the numbers of votes are quite different for different voting methods and the orders are not the same. Additionally, although it is not connected with the subject of the paper, we can notice that the number of minuses is significantly greater than number of pluses in the categorization method. It implies that the balance is positive for one politician only. The categorization method (and also difference between number of votes in approval and disapproval votings) demonstrates a big disappointment of the society with the political class. Next, we shall analyze the results within the neutral, positive and negative scopes of valuation.

Table 1. Results of poll, March 2017

	Maj.	App.	Disapp.	Cat.			
				–	0	+	Balance
AD	192	244	373	550	157	346	–204
JG	5	49	269	645	302	106	–539
JK	17	64	513	786	157	110	–676
MK	5	8	339	704	315	34	–670
BK	50	248	248	440	221	392	–48
JKM	35	79	458	759	181	113	–646
WKK	23	86	175	485	406	162	–323
PK	87	175	325	540	245	268	–272
BN	22	113	206	503	357	192	–311
MO	16	67	324	626	289	138	–488
JP	12	46	369	735	224	95	–640
RP	28	118	256	588	283	182	–406
ARZ	53	164	216	489	306	258	–231
GS	15	85	250	573	315	165	–408
DT	239	379	283	407	178	468	61
AZ	24	85	222	525	382	146	–379
ZZ	3	39	383	719	247	87	–632
other	56	80	29				
Do not vote/do not know	171	152	129				
total	1053						

3 Kendall's Statistical Rank Test

Statistical rank tests are used in situation were we want to compare two orders. The best known are Spearman's rho and Kendall's tau ([14–16]). Kendall's tau is regarded as more precise. We shortly present Kendall's tau below.

Let us consider two random variables X and Y. (x_i, y_i) is a realization of (X, Y). Now, we define $(x_i, y_{i)}$ and (x_j, y_j) are concordant if $x_i > [<]x_j$ and $y_i > [<]y_J$.

So, two pairs are concordant if they are ordered in the same way. Otherwise, they are discordant. $(x_i, y_{i)}$ and (x_j, y_j) are discordant if $x_i > [<]x_j$ and $y_i < [>]y_i$. Then we compute Kendall's τ as follows.

τ = (number of concordant pairs-number of discordant pairs)/[n(n–1)/2].

It is a difference between probabilities that variables give the same and the reverse order.

Another rank statistical ranking method is Spearman's method. It is the Pearson coefficient computed for rank variables. It shows a monotonic, even nonlinear, dependence ([16]).

4 Calculation Neutral Votes

We begin by comparing the result of votings within the neutral scope of valuation. Such scope can be treated as measuring significance and be less sensitive to the voters' emotional reaction. Thus it may be the most suitable to analyze formal properties.

In Table 2 neutrality results are presented. One method and one score of defining neutral votes are compared. The number of neutral votes (0) in the categorization method (column 3) and the number of neutral votes in approval–disapproval method (total number of votes minus the sum of approval and disapproval votes, the last column). For each way of calculating neutral votes the results are ordered from the lowest to the highest.

Table 2. Results of neutral votings

No.	Candidate	0 votes in categorization	No.	Candidates in total-(app+disapp)	Votes in total - (app+disapp)
1	AD	157	1	DT	391
2	JK	157	2	AD	436
3	DT	178	3	JK	476
4	JMK	181	4	JMK	516
5	BK	221	5	PK	553
6	JP	224	5	BK	557
7	PK	245	7	ZZ	631
8	ZZ	247	8	JP	638
9	RP	283	9	MO	662
10	MO	289	10	ARZ	673
11	JG	302	11	RP	679
12	ARZ	306	12	MK	706
13	MK	315	13	GS	718
14	GS	315	14	JG	735
15	BN	357	15	BN	734
16	AZ	382	16	AZ	746
17	WKK	406	17	WKK	792

The results are not the same but they are similar. The position of specific candidate for categorization method with value 0 does not differ much from the position in a method based on the total number of votes minus the sum of approving and disapproving votes. This difference is not greater than 3. There arises the question about similarity between these orders. The correlations of orders were analyzed using statistical ranking tests, Spearman's and Kendall's. The following results were obtained. In what follows ** denotes significance at level 0.01, * - at level 0.05

Kendall's tau is equal to 0.819**, Spearman's rho is equal to 0.943**. In spite of some differences, the order of candidates is similar for examined ways of calculating neutral votes.

5 Positive and Negative Methods

Table 3 presents the Kendall tau (correlation between place in methods described in a row and in a column). Values which are not significant are presented in brackets, $p = 0.05$.

Table 3. Kendall's tau for votings

	Maj.	App.	Disapp.	Cat.–	Cat.0	Cat.+	Cat. balance
App.	0.756**						
Disapp.	0.111	0.258					
Cat. –	0.421*	0.598**	0.632**				
Cat. 0	−0.237	−0.163	0.583**	0.244			
Cat. +	0.701*	0.952**	0.265	0.603**	−0.155		
Cat. balance	0.568	0.804**	0.426*	0.794**	0.037	0.809**	

The positive voting methods (where voters positively evaluate the candidates – the classical majority voting, approval voting, pluses in the categorization method, balance in the categorical method) are positively correlated. It is the same in case of negative voting methods (where voters negatively evaluated the candidates – the disapproval method, minuses in the categorization method). There is no correlation between the positive and negative voting methods, especially between approval and disapproval votings. These votings may be treated as identical from the mathematical point of view, where disapproval voting is a reverse of approval voting. They are not identical from the psychological point of view ([17]) because voters may not decide whether to vote for or against in case of neutrally evaluated candidates.

In Tables 3 and 4 ranks are in decreasing order for votings Maj., App. Cat. + , Cat. balance (positive valuations) and increasing for Disapp., Cat– (negative valuations). Let us note based on Table 3 that there is a close to 0 correlation between the ranks of candidates in approval and disapproval method (although it is not statistically significant). This may support the thesis that these two methods of voting do not measure the same.

Table 4. Spearman's rho for votings

	Maj.	App.	Disapp.	Cat. –	Cat. +	Cat. balance	Cat. 0
App.	0.910*	1					
Disapp.	0.176	0.417	1				
Cat. –	0.612*	0.806**	0.775**	1			
Cat.+	0.872*	0.991**	0.429	0.811**	1		
Cat. balance	0.568**	0.804**	0.426*	0.794**	0.809**	1	
Cat.0	−0.237	−0.163	0.583**	0.244	−0.155	0.037	1

The results are confirmed by Spearman's rho which is presented in Table 4. The dependence is not so distinct when measured by Kendall's rho.

Table 5 presents the results ordered from the highest to the lowest for all considered methods.

Table 5. Results of positive and negative votings

Number	Maj.	App.	Disapp.	Cat. +	Cat. −	Cat. balance
1	DT	DT	WKK	DT	DT	DT
2	AD	BK	BN	BK	BK	BK
3	PK	AD	ARZ	PK	ARZ	AD
4	ARZ	PK	AZ	AD	WKK	ARZ
5	BK	ARZ	BK	ARZ	BN	PK
6	JKM	RP	GS	BN	AZ	BN
7	RP	BN	RP	RP	PK	WKK
8	AZ	WKK	JG	GS	AD	AZ
9	WKK	AZ	DT	WKK	GS	RP
10	BN	GS	MO	AZ	RP	GS
11	JK	JKM	PK	MO	MO	MO
12	MO	MO	MK	JKM	JG	JG
13	GS	JK	JP	JK	MK	ZZ
14	JP	JG	AD	JG	ZZ	JP
15	JG	JP	ZZ	JP	JP	JKM
16	MK	ZZ	JKM	ZZ	JKM	MK
17	ZZ	MK	JK	MK	JK	JK

Let us study Table 5. Analogously to the results of neutral voting methods the difference in the position of a candidate in all positive votings is never greater than 3. It is the same for negative votings. When comparing negative and positive votings we obtain significant differences.

6 Conclusions

It was demonstrated that the results of votings, considered within the same valuation scope are strongly positively correlated in spite of some differences in order. So, there may different winners, but the order of candidates remains similar. It is possible that this expresses some deeper property of voters' preferences directed by the kind of valuation. The results of votings within opposite scopes are not correlated. So, the difference between votings of positive and negative scope is deeper than between different votings of the same scope. There may be formulated a hypothesis that it is connected with the fact that decision maker treats gains and losses in a different way ([18]).

The research was conducted over a representative sample which ensured randomization of respondents. The research concerns elections very remote in time, so actual emotions do not affect the results strongly. Moreover, the winner is many times a politician from the opposition, which is not symptomatic of Polish public opinion.

A source of such similarities of results need further research in order to establish more information.

References

1. Nurmi, H.: Comparing Voting Systems. D. Reidel Publishing Company, Dordrecht (1987)
2. Hołubiec, J.W., Mercik, J.W.: Inside Voting Procedures, p. 3. Accedo Verlagsgesellschaft, Munich (1994)
3. Ordeshook, P.: Game Theory and Political Theory. Cambridge University Press, Cambridge (1986)
4. Brams, S.J., Fisburn, P.C.: Approval voting. A. Political Sci. Rev. **72**(3), 831–847 (1978)
5. Brams, S., Fisburn, P.C.: Approval Voting. Birkhaser, Boston (1983)
6. The "Wisnumurti Guidelines" for Selecting a Candidate for Secretary-General (1996)
7. Felsenthal, D.S.: On combining approval voting with disapproval voting. Behav. Sci. **34**, 53–60 (1989). https://doi.org/10.1002/bs.383040105
8. Alcantud, J.C., Laruelle, A.: Dis&approval voting: a characterization. Soc. Choice Welf. **43**, 1–10 (2014). https://doi.org/10.1007/s00355-013-0766-7
9. Hillinger, C.: Voting and cardinal aggregation of judgments. Discussion Paper 2004-09, University of Munich (2004)
10. Hillinger, C.: The case for utilitarian voting. Homo Oeconomicus **22**, 295–321 (2005)
11. Malawski, M., Przybyszewski, K., Sosnowska, H.: Cognitive effort of voters under three different voting methods-an experimental study. Badania Operacyjne i Decyzje (Oper. Res. Decis.) **3–4**, 69–79 (2010)
12. Przybyszewski, K., Sosnowska, H.: Approval voting as a polling method. J. Polit. Mark. (Submitted to 2018)
13. Chirumbolo, A.: The relationship between need for cognitive closure and political orientation: the mediating role of authoritarianism. Pers. Individ. Differ. **32**, 603–610 (2002)
14. Abdi, H.: Kendall rank correlation. In: Salking, N.J. (ed.) Encyklopedia of measure and statistics. Sage, Thousand Oaks (CA) (2007)
15. Kendall, M.G.: Rank Correlation Methods. Charles Griffin & Company Limited, London (1948)
16. Spearman, C.: The proof and measurement of association between two things. Am. J. Psychol. **15**, 72–101 (1904)
17. Przybyszewski, K., Sosnowska, H.: Approval voting as a method of prediction in political votings. Case of polish elections. In: Nguyen, N.T., Kowalczyk, R., Mercik, J. (eds.) Transactions on Computational Collective Intelligence XXIII. LNCS, vol. 9760, pp. 17–28. Springer, Heidelberg (2016). https://doi.org/10.1007/978-3-662-52886-0_2
18. Tversky, A., Kahneman, D.: A framing of decisions and psychology of choice. Science **211**, 453–458 (1981)

Remarks on Unrounded Degressively Proportional Allocation

Katarzyna Cegiełka, Piotr Dniestrzański, Janusz Łyko,
and Arkadiusz Maciuk[✉]

Wrocław University of Economics,
Komandorska 118/120, 53-345 Wrocław, Poland
{katarzyna.cegielka,piotr.dniestrzanski,janusz.lyko,
arkadiusz.maciuk}@ue.wroc.pl

Abstract. The Lisbon Treaty has legally endorsed degressiveness as a principle of distributing indivisible goods. Yet the principle has been implemented in executive acts with insufficient precision. As a result, it cannot be unambiguously applied in practice. Therefore many theoretical studies have been conducted aiming at a more precisely defined formulation of the principle so that resulting allocations could be explicitly derived from primary rules. This paper belongs to such research stream. It aims at submitting a formal definition of unrounded degressively proportional distribution.

Keywords: Elections · Fair division · Allocation
Unrounded degressive proportionality · Voting

1 Introduction

The term "degressive" is explained by Oxford Living Dictionaries as "reducing by gradual amounts". It emerges in social and economic problems in several contexts. The concept of degressiveness is commonly associated with tax scales. Degressive taxation in personal income taxes implies that taxpayers with higher incomes do not make a due sacrifice on the basis of equity. In degressive taxation the tax payable increases only at a diminishing rate. In some jurisdictions also a term of degressive depreciation of fixed assets is used. Degressive depreciation is derived from the assumption that wear and tear of an asset, thus the reduction in its value, occurs more rapidly at the beginning of its useful life span than in later periods. As a result, an asset becomes depreciated more quickly than e.g. under the straight-line method. Degressiveness also emerges in the context of the fair division problem. It may happen that more entitled agents forsake a part of their claims to benefit those less entitled, for the good of the common goals. Then a resulting allocation is proportional to modified claims, and degressively proportional to original claims. One practical example of such construct is the apportionment of seats in the European Parliament to Member States of the European Union.

In all those examples degression is an idea defining the general principles in case of taxation, depreciation or division. Nevertheless, this idea must be supported by special, additional legal regulations necessary in practical applications. Such regulations should ensure unambiguous and transparent applications, easily comprehensible by all interested

© Springer-Verlag GmbH Germany, part of Springer Nature 2018
N. T. Nguyen et al. (Eds.): TCCI XXXI, LNCS 11290, pp. 55–63, 2018.
https://doi.org/10.1007/978-3-662-58464-4_6

parties. There are appropriate legal regulations in two former areas, underpinned by suitable theoretical foundations. In case of degressively proportional division of goods the respective factors are still unavailable. Hereinafter some thoughts are presented in order to formalize the term of degressively proportional division. The formalization is accomplished by an unambiguous definition including currently binding legal acts as well as prior applications of the idea of this type of allocation.

2   Rounded Degressive Proportionality

The principle of degressive proportionality was introduced to harmonize the rules of apportionment of seats in the European Parliament to Member States of the European Union. The article 1 point 15 of The Treaty of Lisbon states that "The European Parliament shall be composed of representatives of the Union's citizens. They shall not exceed seven hundred and fifty in number, plus the President. Representation of citizens shall be degressively proportional, with a minimum threshold of six members per Member State. No Member State shall be allocated more than ninety-six seats" [1]. This provision introduces for the first time a name of the form of parliamentary representation of citizens, that is degressively proportional. However, a new term in the area of allocating indivisible goods is not defined in the treaty, merely introduced as an idea of fair distribution. The principles of degressive proportionality have been defined more precisely in the *Report on the composition of the European Parliament* and the *Motion for a European Parliament resolution*, the draft of which is attached to the Report, which reads: "The principle of degressive proportionality means that the ratio between the population and the number of seats of each Member State must vary in relation to their respective populations in such a way that each Member from a more populous Member State represents more citizens than each Member from a less populous Member State and conversely, but also that no less populous Member State has more seats than a more populous Member State" [2].

On this basis one can put forward the definition of degressive proportionality accepted in the subject matter literature, the so-called Rounded Degressive Proportionality (RDP).

Definition 1 (RDP sequence). Let $P = (p_1, p_2, \ldots, p_n)$ denote a positive, nondecreasing sequence. We say that a positive sequence $S = (s_1, s_2, \ldots, s_n)$, $s_i \in \mathbb{N}_+$, is *degressively proportional (RDP) with respect to* P if $s_1 \leq s_2 \leq \ldots \leq s_n$ and $\frac{p_1}{s_1} \leq \frac{p_2}{s_2} \leq \ldots \leq \frac{p_n}{s_n}$.

Definition 1a (RDP allocation). If $\sum_{i=1}^{n} s_i = H$ holds in Definition 1, then a sequence S is called an *RDP allocation of H goods* among agents with claims (values, entitlements) $P = (p_1, p_2, \ldots, p_n)$.

As regards the European Parliament, respective denotations in Definition 1a are evident: p_i – the population of the ith Member State, and s_i – the number of its seats. Generally, the sequence P represents claims or entitlements of agents. The remaining constraints imposed by the Lisbon Treaty on the minimum, maximum and total number of seats in the European Parliament: $s_i \geq 6, s_n \leq 96$, and $H = \sum_{i=1}^{n} s_i \leq 751$, are sometimes called the *boundary conditions* and formulated as inequalities, but practically all researchers in this area assume that they are equalities, i.e. $s_i = 6, s_n = 96$ and $H = 751$.

The methods determining allocations, which satisfy the conditions of Definition 1a, are presented among others in the papers [3–5]. The main idea of these methods consists in recursive construction of respective allocations or in searching the complete solution space, i.e. all degressively proportional allocations, which satisfy the boundary conditions, and in selecting one allocation out of them, subject to a predetermined, additional criterion. In some cases it can be impossible to determine an RDP allocation, while in other cases the main idea of fair and rational division can be significantly distorted.

Table 1. Influence of the structure of claims on potential RDP allocations.

A	B	C	D	E
P	RDP	RDP	P	RDP
	H = 5	H = 11		H = 11
3	1	2	3	1
5	1	2	4	1
9	1	2	9	2
15	1	2	15	3
23	1	3	23	4

One can observe e.g. that the claims of agents shown in column A of Table 1 do not generate any RDP allocation, if the number of goods to be allocated is $H = 6$. With $H = 5$, there is exactly one RDP allocation – shown in column B of Table 1, but it does not reflect the structure of claims in any way. What's more, proceeding in a similar way one can point to a case, where the ratio between the claims of the greatest and the smallest agent is arbitrarily large, and the single RDP allocation is the equal division. The minimum number of goods allowing for some RDP allocation different from the equal division, is $H = 11$. As easily seen, the only RDP allocation, given the claims from column A, is shown in column C, and is quite undiversified. Yet a relatively small modification of the sequence P (column D) results in a significantly more diversified allocation (column C and E), given the same number of goods to be allocated. This proves a considerable sensitivity of RDP allocations to changes in the structure of the sequence of claims.

The mentioned weaknesses of RDP allocations call for more effort in research reported by the subject matter literature and in practical applications, to find different solutions, which approach the idea of degressive proportionality in a less restrictive manner. This research stream allows distributions, which do not satisfy the second condition included in Definition 1. They are called Unrounded Degressive Proportionality (UDP) allocations. From a theoretical point of view, the problem is created by the deficiency of a strict definition of the idea of UDP, both in legal acts and in the literature. In practice, this deficiency can result in problems with deciding whether a given allocation, that is not an RDP allocation, can be treated as a UDP allocation. The necessity of more precise formulations related to degressively proportional allocation was noticed as early as in 2007, when Lamassoure and Severin stated in their report that "ideal alternative would be to agree on an undisputed mathematical formula of 'degressive proportionality' that would ensure a solution not only for the present revision but for future enlargements or modifications due to demographic changes" [2].

3 Unrounded Degressive Proportionality

The first attempt to legally endorse such allocations, which do not have to satisfy the second condition in Definition 1 was made in 2011 during the session of the Committee on Constitutional Affairs of the European Parliament. It was the accomplishment of the postulate to formulate the principle of degressive proportionality in such a way, that any allocation generated under the method known as Cambridge Compromise, could be deemed as satisfying the conditions of the Lisbon Treaty [6]. The change consisted in requiring that the judicial interpretation: "the principle of degressive proportionality means that the ratio between the population and the number of seats of each Member State must vary in relation to their respective populations in such a way that each Member from a more populous Member State represents more citizens than each Member from a less populous Member State and conversely, but also that no less populous Member State has more seats than a more populous Member State" [2], were replaced by the following: "the principle of degressive proportionality means that the ratio between the population and the number of seats of each Member State **before rounding to whole numbers** must vary in relation to their respective populations in such a way that each Member from a more populous Member State represents more citizens than each Member from a less populous Member State and conversely, but also that no less populous Member State has more seats than a more populous Member State" [6].

Such approach is naturally justified in the light of integer proportional allocations. In those cases, the so-called quotas of allocation, which are exact realizations of the principle of proportionality, in general are not integer, and have to be replaced by some approximations, e.g. provided by divisor methods. In other words, the generated allocation is strictly proportional only before rounding to integers. This is why later modifications of our understanding the principle of degressive proportionality have been called UDP. Anyway, the problem persisted in the unclear understanding of the phrase "before rounding to whole numbers" (see Table 2).

The above-mentioned method, called Cambridge Compromise, develops prior proposals of shifted proportionality put forward by [7, 8]. The method applies the allocation function[1] $A(p) = \min\{b + \frac{p}{d}, M\}$, where p denotes the size of population in a State, and the total number of seats assigned to Member States amounts to

---

[1] Detailed mathematical analysis of the allocation functions the reader will find in [9].

Table 2. The composition of the European Parliament by various methods and various type of rounding.

	Population	AFCO	CC	Parabolic			Base+Power		
				Down	Middle	Up	Down	Middle	Up
Germany	82064489	96	96	96	96	96	96	96	96
France	66661621	79	91	84	83	83	81	81	81
Italy	61302519	76	84	79	78	78	76	76	76
Spain	46438422	59	65	64	64	62	61	61	61
Poland	37967209	52	54	55	54	53	52	52	52
Romania	19759968	33	31	33	32	32	32	32	31
Netherlands	17235349	29	28	29	29	28	29	29	29
Belgium	11289853	21	20	21	21	21	22	22	21
Greece	10793526	21	19	21	21	20	21	21	21
Czech Rep.	10445783	21	19	20	20	20	21	21	20
Portugal	10341330	21	19	20	20	20	21	21	20
Sweden	9998000	21	18	19	19	19	20	20	20
Hungary	9830485	21	18	19	19	19	20	20	20
Austria	8711500	19	17	18	18	18	18	18	18
Bulgaria	7153784	17	15	15	16	16	16	16	16
Denmark	5700917	14	13	13	14	14	14	14	14
Finland	5465408	14	13	13	13	13	14	14	14
Slovakia	5407910	14	12	13	13	13	14	14	14
Ireland	4664156	13	11	12	12	12	13	13	13
Croatia	4190669	12	11	11	11	12	12	12	12
Lithuania	2888558	11	9	9	10	10	10	10	10
Slovenia	2064188	8	8	8	8	9	9	9	9
Latvia	1968957	8	8	8	8	9	8	8	9
Estonia	1315944	7	7	7	7	8	7	7	8
Cyprus	848319	6	7	6	7	7	6	6	7
Luxembourg	576249	6	6	6	6	7	6	6	7
Malta	434403	6	6	6	6	6	6	6	6
EU-27	445519516	705	705	705	705	705	705	705	705

Legend. AFCO: The composition approved by the Committee on Constitutional Affairs [13]; CC – Cambridge Compromise, Parabolic and Base+Power – see [9].

$T(d) = [\![A(p_1)]\!] + [\![A(p_2)]\!] + \ldots + [\![A(p_1)]\!]$, where $[\![x]\!]$ denotes the rounding of number x, and divisor d is selected to satisfy $T(d) = H$.

One of the first proposals of degressively proportional allocations, where degressive proportionality appears only prior to rounding, was presented by [10]. The author put forward the allocation function of the form $A(p) = p^u$, where u is a fixed parameter from the interval $(0, 1)$ to determine the allocation of seats in the European Parliament[2]. The values $A(p_i)$ for respective coordinates of the sequence P create a new sequence $P' = (A(p_1), A(p_2), \ldots, A(p_n))$, which is a basis of proportional division of a given number of goods. The value of parameter u can be determined arbitrarily or agreed upon by participants of the distribution.

A similar idea also underpins a parabolic method, which applies a quadratic allocation function $A(p) = ap^2 + bp + c$. The coefficients of the function A are chosen so as to satisfy the boundary conditions of the allocation, i.e. any requirements concerning minimum and maximum numbers of goods to be allocated to individual agents. The number of goods s_i allocated to the ith agent amounts to the value of the function A rounded to the nearest integer. The parabolic method was proposed to determine the allocation of seats in the European Parliament. Its representatives are quoted as saying that "among the various possible mathematical formulae for implementing the principle of degressive proportionality, the 'parabolic' method is one of the most degressive" [12].

The mentioned proposals to realize degressively proportional allocations indicate that their authors comprehend a degressively proportional division (UDP) in a following way:

Definition 2 (UDP sequence). Let $P = (p_1, p_2, \ldots, p_n)$ denote a positive, non-decreasing sequence. We say that a positive sequence $S = (s_1, s_2, \ldots, s_n)$, $s_i \in \mathbb{N}_+$, is *degressively proportional (UDP) with respect to P* if $s_1 \leq s_2 \leq \ldots \leq s_n$ and there exists a rounding rule $[\![.]\!]$ and a function $A : [p_1, p_n] \to \mathbb{R}_+$ such that $t_1/A(t_1) \leq t_2/A(t_2)$ for $t_1 \leq t_2$ and $(s_1, s_2, \ldots, s_n) = ([\![A(p_1)]\!], [\![A(p_2)]\!], \ldots, [\![A(p_n)]\!])$.

Definition 2a (UDP allocation). If $\sum_{i=1}^{n} s_i = H$ holds in Definition 2, then the sequence S is called an *UDP allocation* of H goods among agents with given claims (value, entitlements) P.

The application of the allocation function in the definition of UDP makes it easier to interpret any proposed solutions. For instance, Pukelsheim's parabolic method allows to estimate the decrease of the ratio between the number of seats in the European Parliament and the population in a given Member States of the European Union, as populations increase. The method proposed by Haman allows to control the degression rate of allocations. On the other hand though, approaching UDP through the allocation function may create some limitations. Proposed allocations are necessarily connected with the choice of a certain class of functions, and in consequence a wider review of potential solutions is not possible. We propose a different definition of UDP, which similarly as divisor methods in proportional allocations, will not require an allocation function.

---

[2] Haman argues that the value of parameter u can be then interpreted as a measure of degression of allocations. A smaller value of this parameter indicates more degression of the allocation. The smaller the value u, the more degressive allocation (closer to equal division). A given allocation is less degressive (closer to a proportional allocation), as u approaches unity. Haman continues the research into degressive proportionality in a weak case (UDP) and in a strong case (RDP) [11].

Definition 3 (UDP allocation). Let $P = (p_1, p_2, \ldots, p_n)$ denote a positive, non-decreasing sequence. We say that a positive sequence $S = (s_1, s_2, \ldots, s_n), s_i \in \mathbb{N}_+$, is called an *UDP degressively proportional of H goods* among agents with given claims (values, entitlements) P if $s_1 \leq s_2 \leq \ldots \leq s_n$ and there exists a positive, non-decreasing sequence $P' = (p'_1, p'_2, \ldots, p'_n)$ such that $\frac{p_1}{p'_1} \leq \frac{p_2}{p'_2} \leq \ldots \leq \frac{p_n}{p'_n}$, and S is a proportional allocation with respect to P' of H goods obtained with a divisor method with a fixed rounding rule.

Proposition 1. Definition 2a is equivalent to Definition 3.

Proof. In order to demonstrate that Definition 2a implies Definition 3, it suffices to take $p'_i = A(p_i), d = 1$. Then $\frac{p_1}{p'_1} \leq \frac{p_2}{p'_2} \leq \ldots \leq \frac{p_n}{p'_n}$ and $S = \left(\left[\!\left[\frac{p'_1}{d} \right]\!\right], \left[\!\left[\frac{p'_2}{d} \right]\!\right], \ldots, \left[\!\left[\frac{p'_n}{d} \right]\!\right] \right) = (\llbracket A(p_1) \rrbracket, \llbracket A(p_2) \rrbracket, \ldots, \llbracket A(p_n) \rrbracket)$ is an allocation we seek after, where $[.]$ denotes a fixed rounding rule. The converse implication results from the substitution $A(p_i) = \frac{p'_i}{d}$, where d is a divisor generating the UDP according to Definition 3. Then we have $\frac{p_i}{A(p_i)} = \frac{p_i}{\frac{p'_i}{d}} = \frac{dp_i}{p'_i}$. Because $d > 0$, and by assumption that $\frac{p_1}{p'_1} \leq \frac{p_2}{p'_2} \leq \ldots \leq \frac{p_n}{p'_n}$, yields $\frac{dp_1}{p'_1} \leq \frac{dp_2}{p'_2} \leq \ldots \leq \frac{dp_n}{p'_n}$. Hence $\frac{p_1}{A(p_1)} \leq \frac{p_2}{A(p_2)} \leq \ldots \leq \frac{p_n}{A(p_n)}$. Therefore $\left(\left[\!\left[\frac{p'_1}{d} \right]\!\right], \left[\!\left[\frac{p'_2}{d} \right]\!\right], \ldots, \left[\!\left[\frac{p'_n}{d} \right]\!\right] \right) = S$ is the UDP with respect to P, as understood in the sense of Definition 2a. The equality $A(p_i) = p'_i/d$ defines the allocation function A at points $p_1, p_2, \ldots, p_n$. To obtain the function A in the entire interval $[p_1, p_n]$, it suffices to take a suitable piecewise linear function. ∎

Definitions 2, 2a and 3 are very general because of the too general term of rounding rule. It suffices to consider Definition 4 given by [14] in order to notice that resulting allocations can be very distant from a degressively proportional allocation, which is perceived by intuition.

Definition 4 (rounding rule). A *jumppoint sequence* $y(0), y(1), y(2), \ldots$ is an unbounded sequence satisfying $y(0) = 0 \leq y(1) < y(2) < \ldots$. A jumppoint sequence defines a *rounding rule* $[.]$ by setting, for all $x \geq 0$ and $n \in \mathbb{N}$,

$$\llbracket x \rrbracket = \begin{cases} \{0\} & \text{in case } x = 0 \\ \{n\} & \text{in case } x \in (y(n), y(n+1)) \\ \{n-1, n\} & \text{in case } x = y(n) > 0. \end{cases}$$

Definition 4 offers freedom of choice when rounding. Suitably specifying jumppoint sequence we can get roundings, which are inconsistent with a common understanding of the term. For instance, if we have initial terms of a jumppoint sequence equal $y(0) = 0, y(1) = 3, y(2) = 4, y(3) = 7, y(4) = 21, y(5) = 27$ then we get $\llbracket 6 \rrbracket = 2$ (because $6 \in [y(2), y(3)]$) and $\llbracket 15.9 \rrbracket = 3$ (because $15.9 \in [y(3), y(4)]$).

It turns out that if we comprehend rounding as being consistent with Definition 4, then assuming additionally that the claims in sequence P have different values, any sequence S satisfying $s_1 \leq s_2 \leq \ldots \leq s_n$ can be considered a UDP allocation with respect to the sequence P.

Proposition 2. For fixed $H > 0$ and a positive, non-decreasing sequence $P = (p_1, p_2, \ldots, p_n)$ such that $p_i \neq p_j$ for $i \neq j$, every positive, non-decreasing sequence $S = (s_1, s_2, \ldots, s_n)$, $s_i \in \mathbb{N}_+$, $H = \sum_{i=1}^{n} s_i$ is UDP allocation with respect to P.

Proof. Let $P = (p_1, p_2, \ldots, p_n)$ and satisfy the assumption of Proposition 2. We will demonstrate that the sequence S is UDP allocation with respect to P, where UDP is understood to be consistent with Definition 2a.

We have to prove that there exists a non-decreasing sequence $P' = (p_1', p_2', \ldots, p_n')$ such that $\frac{p_1}{p_1'} \leq \frac{p_2}{p_2'} \leq \ldots \leq \frac{p_n}{p_n'}$, and a jumppoint sequence $y(0), y(1), \ldots$ such that the equalities $[\![p_i']\!] = s_i$ hold for every $i = 1, 2, \ldots, n$. It suffices to take $P' = P, d = 1$ and any jumppoint sequence $y(0), y(1), \ldots$ such that $y(s_i) < p_i < y(s_i + 1)$ hold for every $i = 1, 2, \ldots, n$. ∎

Proposition 2 demonstrates that defining UDP consistent with Definition 2a or equivalent Definition 3 isn't appropriate, due to excessively general definition of rounding, with a common-sense understanding of degressive proportionality and with its practical applications developed so far. Considering the accomplishments related to divisor methods employed in proportional distribution, understanding their properties, and widespread practical applications, the following definition of UDP allocation is put forward.

Definition 5 (UDP allocation). Let $P = (p_1, p_2, \ldots, p_n)$ denote a positive, non-decreasing sequence. We say that a positive sequence $S = (s_1, s_2, \ldots, s_n)$, $s_i \in \mathbb{N}_+$, is called an *UDP allocation of H goods* among agents with given claims (value, entitlements) P if $s_1 \leq s_2 \leq \ldots \leq s_n$ and there exists a positive, non-decreasing sequence $P' = (p_1', p_2', \ldots, p_3')$ such that $\frac{p_1}{p_1'} \leq \frac{p_2}{p_2'} \leq \ldots \leq \frac{p_n}{p_n'}$ and S is a proportional allocation with respect to P' of H goods obtained with a rounding upwards, downwards or to the nearest integer.

4 Conclusions

Definition 5 puts forward a new interpretation of the term of unrounded degressive proportionality, which extremely reduces a range of allocations potentially deemed as UDP. All proposals known from the literature of practical applications of the idea of degressively proportional distribution of goods, which make use of allocation functions and standard rounding downwards, upwards and to the nearest integer, are obviously UDP in the sense of Definition 5, in view of demonstrated equivalence of Definitions 2a and 3. Besides narrowing the concept of rounding rule, Definition 5 shifts emphasis from allocation functions on sequences P', which can be interpreted as modifiers of entitlements claimed by individual agents. In practical applications typically concerning a fixed sequence P, there is no need to adjust claims in the entire interval $[p_1, p_n]$, as it is indirectly accomplished by means of allocation function. Now it suffices to modify claims at points p_i.

Shifting emphasis on adjustment of claims is in agreement with the idea of degressive proportionality and solidarity, when larger agents, with greater claims, share their entitlements with smaller agents. This is an example of manifesting European solidarity [2], which encourages attitudes when "the more populous States agree to be under-represented in order to allow the less populous States to be represented better". In addition, Definition 5 explicitly relates to methods of proportional allocation of indivisible goods, and as a result, the considerable theoretical achievements in this field can be employed. Defining UDP allocations itself does not obviously remove a problem of their unambiguity. Still no single, particular allocation is designated. Given the sequence P, there may be a number of sequences P', but also one particular sequence P', resulting from the use of a divisor method and various feasible roundings may specify various integer allocations S.

References

1. Treaty of Lisbon. Official J. Eur. Union. C 306, **50** (2007)
2. Lamassoure, A., Severin, A.: Report on the composition of the European Parliament. Committee on Constitutional Affairs, A6-0351 (2007)
3. Maciuk, A.: Przykłady podziałów mandatów do Parlamentu Europejskiego spełniających warunki Traktatu Lizbońskiego [Examples of the divisions of seats to the European Parliament that meet the conditions of the Treaty of Lisbon]. Prace Naukowe Uniwersytetu Ekonomicznego we Wrocławiu. Ekonometria **33**(198), 85–96 (2011)
4. Florek, J.: A numerical method to determine a degressive proportional distribution of seats in the European Parliament. Math. Soc. Sci. **63**(2), 121–129 (2012)
5. Łyko, J., Rudek, R.: A fast exact algorithm for the allocation of seats for the EU Parliament. Expert Syst. Appl. **40**, 5284–5291 (2013)
6. Grimmett, G.R.: European apportionment via the Cambridge Compromise. Math. Soc. Sci. **63**(2), 68–73 (2012)
7. Pukelsheim, F.: A parliament of degressive representativeness? Institut für Mathematik, Universität Augsburg, Preprint Nr 015/2007 (2007)
8. Pukelsheim, F.: Putting citizens first: representation and power in the European Union. In: Cichocki, M., Życzkowski, K. (eds.) Institutional Design and Voting Power in the European Union, pp. 235–253. Ashgate, London (2010)
9. Słomczyński, W., Życzkowski, K.: Mathematical aspects of degressive proportionality. Math. Soc. Sci. **63**(2), 94–101 (2012)
10. Haman, J.: Degresywnie proporcjonalny podział mandatów w Parlamencie Europejskim [Degressively proportional division of seats in the European Parliament]. Decyzje **8**, 53–78 (2007)
11. Haman, J.: The concept of degressive and progressive proportionality and its normative and descriptive applications. Studies of Logic, Grammar and Rhetoric (2017). https://doi.org/10.1515/slgr-2017-0019. (in print)
12. Gualtieri, R., Trzaskowski, R.: Report on the composition of the European Parliament with a view to the 2014 elections. A7-0041/2013 (2013)
13. Report on the composition of the European Parliament (28 January 2018) (2017/2054(INL) – 2017/0900(NLE)). http://www.europarl.europa.eu
14. Pukelsheim, F.: Proportional Representation. Apportionment Methods and Their Applications. Springer, Cham (2014)

On Measurement of Control in Corporate Structures

Jacek Mercik[1] ⓘ and Izabella Stach[2(✉)] ⓘ

[1] WSB University in Wroclaw, Fabryczna 29/31, 53-609 Wroclaw, Poland
jacek.mercik@wsb.wroclaw.pl
[2] AGH University of Science and Technology,
Al. Mickiewicza 30, 30-059 Krakow, Poland
istach@zarz.agh.edu.pl

Abstract. Various methods to measure power control of firms in corporate shareholding structures are introduced. This paper is a study of some game-theoretical approaches for measuring direct and indirect control in complex corporate networks. We concentrate on the comparison of methods that use power indices to evaluate the control-power of firms involved in complex corporate networks. More precisely, we only thoroughly analyze the Karos and Peters and the Mercik and Lobos approaches. In particular, we consider the rankings of firms given by the considered methods and the meaning of the values assigned by the power indices to the stock companies presented in corporate networks. Some new results have been obtained. Specifically, taking into account a theoretical example of a corporate shareholding structure, we observe the different rankings of investors and stock companies given by the Φ index introduced by Karos and Peters in 2015 and the implicit index introduced by Mercik and Lobos in 2016. Then, some brief considerations about the reasonable requirements for indirect control measurement are provided, and some ideas of modifying the implicit index are undertaken. The paper also provides a short review of the literature of the game-theoretical approaches to measure control power in corporate networks.

Keywords: Corporate control · Direct and indirect control
Cooperative game theory · Power indices

1 Introduction

In corporate shareholding networks, firms exercise control over each other by owning each other's stocks. Control can be direct and indirect. Indirect control is exerted across pyramid linkages, circular cross-ownership (hereafter called "loops"), and cross-shared holdings. By a cross-shareholding (or more simply, cross-holding), we mean the simple cross-ownership situation where Firm X has voting rights in Firm Y and, at the same time Firm Y has voting rights in Firm X. In a loop structure, we have at least three firms involved. These mechanisms (top-down ownerships, cross-holdings, and loops) enable firms to exert indirect control over firms with low cash-flow rights. In complex shareholding structures with many firms, loops, and cross-shareholdings, it is hard to

© Springer-Verlag GmbH Germany, part of Springer Nature 2018
N. T. Nguyen et al. (Eds.): TCCI XXXI, LNCS 11290, pp. 64–79, 2018.
https://doi.org/10.1007/978-3-662-58464-4_7

detect mutual relationship loops. More precisely, it is difficult to find ultimate share-holders or coalitions of firms that can exert control over a particular stock company or even over all companies in the network. Thus, indirect control in shareholding networks have become an interesting issue to investigate. Then, as shareholding size is not a good measure of the control power of an investor's company in a shareholding network, many various methods to model and measure the indirect control have been introduced. For example, Claessens, Djankov, and Lang (in [1]) and Faccio and Lang (in [2]) analyzed the ownership and control of corporations in East Asian and Westerns European countries, respectively. They investigated thousands of corporations and obtained interesting findings about real-world corporations. They used non-game-theoretical approaches to measure shareholder control power. More specifically, they used the chain-based methods (for further information, see also [3]). In this paper, we place interest only on those methods that use the cooperative game theory framework. Precisely, we are interested in those that use power indices to measure the control power of firms.

The first and rather pioneering game-theoretical approach to measure the indirect control in the corporate networks was introduced by Gambarelli and Owen in 1994 (see [4]). In the following years, numerous other models were introduced by scholars. Let us list some of them in chronological order (i.e., as these approaches appeared in the literature): Turnovec [5], Denti and Prati [6, 7], Hu and Shapley [8, 9], Crama and Leruth [10], Levy [11, 12], Karos and Peters [13], Mercik and Lobos [14], and Levy and Szafarz [3]. There are some similarities between some of these methods. For example, some of them reference the same indices, or there are some resemblances in the procedures for taking indirect control into account. Some of the methods successfully measure the indirect control of firms in acyclic corporate networks; however, they fail to measure the control power of agents in networks with cyclic ownerships. Some discussions on this matter (i.e., on comparing the approaches) can be found in Bertini, Mercik, and Stach [15] and also in Levy and Szafarz [3]. In Kołodziej and Stach [16] and Stach [17], some simulations were performed in several examples of shareholding structures, taking into account the Denti and Prati [6] approach and four power indices: the Shapley and Shubik [18], Banzhaf [19], Johnston [20], and Holler [21] indices. The latest method introduced in 2017 by Levy and Szafarz [3] follows the Crama and Leruth [10] approach. The Levy and Szafarz approach combines the use of the Banzhaf [19] power index with Markov voting chains. Only three of the existing methods in the literature measure the control power of not only investors (firms without shareholders) but also of companies. Namely, these methods are those of Karos and Peters [13], Mercik and Lobos [14], and Levy and Szafarz [3]. The good and critical literature reviews on game-theoretical approaches can be found in Crama and Leruth [22], Karos and Peters [13], Levy and Szafarz [3], and Stach [17].

In this paper, we focus on two approaches: those proposed by Karos and Peters [13] and Mercik and Lobos [14]. These two methods of measuring control power take into account all firms involved in shareholding networks. The numerical results are difficult to compare, as the power indices proposed by these methods have different ranges (for example). However, we can compare the rankings of firms provided by these methods. Surprisingly, using an example of a theoretical corporate shareholding structure, we find that the rankings are different. Therefore, we try to investigate what exactly these methods measure. Then, we regard an important issue; that is to say, the aspect of properties that should satisfy a good measure of indirect control power. Karos and Peters [13] just started their research from a set of axioms and then arrived to their uniquely defined power index. A short discussion on this issue will be undertaken in Sect. 5. This work can be seen as a continuation of the research started in [15] and carried on in [16, 17].

The rest of the paper is structured as follows. In Sect. 2, the basic definitions and notations of cooperative game theory are presented. In Sect. 3, we present an example of a theoretical shareholding structure. Section 4 compares the Karos and Peters [13] and Mercik and Lobos [14] approaches in the theoretical example and provides the main result that these methods are not qualitatively equivalent. Then, the interpretations of the values assigned to stock companies by the methods are also discussed. Section 5 is devoted to the required properties of indirect control measurement. Here as well, a proposition of modifying the implicit index [14] is mentioned. Finally, Sect. 6 concludes with some remarks and some lines for further development.

2   Basic Definitions and Notations

A cooperative n-person game is a pair (N, v) where $N = \{1, 2, \ldots, n\}$ denotes a finite set of players and $v : 2^N \to R$ is the *characteristic function* on the 2^N family of all subsets of N, with $v(\emptyset) = 0$. We use the $|\cdot|$ operator to denote the cardinality of a finite set. All subsets S of N are called coalitions, and $v(S)$ is the worth of coalition S in game v. Cooperative game (N, v) will be referred to simply as v in this paper. Cooperative game v is *monotonic* if $v(S) \leq v(T)$ holds for all $S \subset T \subseteq N$. A *simple* game is a monotonic game v if v takes only two values: $v(S) = 0$ for *losing coalition* $S \subseteq N$ and $v(S) = 1$ for *wining coalition* $S \subseteq N$. By $W(v)$, we denote the set of all winning coalitions in game v. Player i is *critical* in winning coalition S if $v(S \setminus \{i\}) = 0$. Coalition S is called a *vulnerable* coalition if there is at least one critical player among its members. By VC, we denote the set of all vulnerable coalitions. Winning coalition S is called a *minimal winning coalition* if all of its members are critical. By $W^m(v)$, we denote the set of all minimal winning coalitions in v. Every simple game can be defined by either W or equivalently by W^m. For any player i in N, $C_i = \{S \subseteq N : i \text{ is critical in } S\}$.

Simple game $v = [q; w_1, \ldots, w_n]$ is a *weighted majority game* when (i) $\sum_{i \in N} w_i = 1$, weight $w_i \geq 0$ for all $i \in N$, and $w(S) = \sum_{i \in S} w_i$ represents weight of coalition S; (ii) threshold $q > 0$ (called also majority quota) establishes winning coalitions and $\sum_{i \in N} w_i \geq q$; (iii) $v(S) = 1$ if $\sum_{i \in S} w_i > q$; otherwise, $v(S) = 0$. Often, weighted majority games are used to model voting situations, and the weights can represent the voting rights owned in a company in a corporate shareholder structure (for example).

Power index f on the sets of all simple games is a map that assigns unique vector $f(v) = (f_1(v), f_2(v), \ldots, f_n(v))$ to every simple game v, where $f_i(v)$ represents the payoff to Player i and can be interpret as a measure of the influence of Player i in v. The best-known power index is the Shapley-Shubik index [18], which is defined for every simple game v and $i \in N$ by

$$\sigma_i(v) = \sum_{S \in C_i} \frac{(s-1)!(n-s)!}{n!},$$

where $s = |S|$. For further information of σ, see [23], for example.

The most-used power indices by researchers to measure the control power of firms in shareholding structures are the Shapley-Shubik and Banzhaf indices. The absolute Banzhaf index, β, was introduced by Banzhaf [19] for any v and $i \in N$ and is defined as:

$$\beta_i(v) = \frac{|C_i|}{2^{n-1}}.$$

For more explanations, see [24], for example.

The approach by Mercik and Lobos [14] uses a modification of the Johnston power index. The Johnston index, γ, was introduced by Johnston in [20]. For any $S \in SV$ in v, $r(S)$ represents the reciprocal of the number of critical players in S. Let us define $r_i(S)$ as follows: if i is critical in S, then $r_i(S) = r(S)$; otherwise, $r_i(S) = 0$. The absolute Johnston index for any v and $i \in N$ is defined as $\bar{\gamma}_i(S) = \sum_{S \in VC} r_i(S)$, and the Johnston index is as follows: $\gamma_i(S) = \frac{\sum_{S \in VC} r_i(S)}{\sum_{i=1}^{n} \sum_{S \in VC} r_i(S)}$.

In regard to power indices, there are some desirable properties. Let us briefly and informally present some of them. The reader can find formal versions of these properties in [25–29]. The *efficiency property* states that the players' power values add up to 1 for all players. The *non-negativity property* states that the players' power value should not be negative. The *null player property* (or *dummy property*) states that a player who is not critical in any coalition has a measure of power equal to zero. The *null player removable property* states that, after removing the null players from a game, the non-null players' measures of power remain unchanged. The *symmetry property* (or *anonymity postulate*) states that a player's power value should not depend on his name. Therefore, the "symmetric" players should have equal power. The *transfer property* states that the elimination of a minimal coalition from the set of winning coalitions has the same effect on the power measure of every game in which this coalition is minimal winning. The *bloc property* states that players should have an advantage from the formation of a coalition. Thus, an alliance between two players should result in a greater power value than the power of a singular player. *Dominance property* states that, in a weighted voting body, a player with a larger voting weight cannot receive less power than a player with a smaller voting weight. *Donation property* says that, in a weighted voting body, a player cannot gain power by donating some of her or his voting weight to another player.

By investor or ultimate company, we mean a firm that is not directly control by any firm in a corporate network (or in other words, a firm that does not have any shareholders).

3 Example of Shareholding Network

Let us consider an example of the cyclic shareholding structure that appeared in [17] and was called the theoretical example. Table 1 presents the direct ownership in this example, and Fig. 1 represents the direct ownerships in a graphical way. The values in Table 1 and on the arcs in Fig. 1 represent (in percentages) the voting rights that one firm has in a company. This example is rather complex. In this structure, there are 13 firms, 5 stock companies (1, 2, 3, 4, 5), and 7 investors (6, 7, 8, 9, 10, 11, 12, 13). In Fig. 1, we can see that there are some loops and a cross-shareholding in the network. For example, Company 2 holds a 30% stake in Company 3, which has a 10% stake in Company 5, and Company 5 holds a 25% stake in Company 2. Then, Company 1 has 3% of the voting rights in Company 4, and Company 4 holds 30% of the voting rights in Company 1. In [17] (where this example was deeply studied), some simulations were performed and some power indices were calculated in order to measure the power control of the agents in the network. However, in this paper, we only apply the Karos and Peters [13] and Mercik and Lobos [14] approaches to this example (see Sects. 4.1 and 4.2).

Table 1. Matrix representation of theoretical example.

Firm	1	2	3	4	5
1				3%	
10		35%			
11	5%			2%	
12					45%
13	5%				
2	35%		30%		
3	20%				10%
4	30%	5%			
5		25%			
6		5%	30%		15%
7		30%	40%	90%	30%
8				5%	
9	5%				

In order to assess the control power of firms in this example, we consider a simple majority rule; i.e., majority quota $q = 50\%$ describing the necessary voting rights (in percentage) to make a proposal pass.

For each of the stock companies in the shareholding structure presented in Fig. 1, Table 2 presents the sets of minimal winning coalitions and vulnerable coalitions. The minimal winning coalitions for this example were just given in [17]. The sets of vulnerable coalitions for each company in Table 2 were found to take only direct ownerships, whereas the sets of minimal winning coalitions are given to taking both direct and indirect control under consideration. The minimal winning coalitions and vulnerable coalitions are crucial for calculations respective of the Φ and implicit indices.

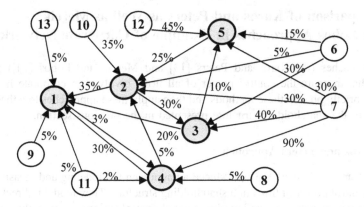

Fig. 1. Theoretical shareholding structure.

Table 2. Minimal winning and vulnerable coalitions in different target companies.

Company	Minimal winning coalitions considering direct and indirect control	Vulnerable coalitions considering only direct control. Critical players are underscored
1	{2, 3}, {2, 4}, {2, 6}, {2, 7}, {5, 7}, {6, 7}, {7, 10}, {7, 12}, {3, 10, 12}, {3, 5, 10}, {3, 4, 9}, {3, 4, 11}, {3, 4, 13}, {3, 7, 9}, {3, 7, 11}, {3, 7, 13}, {4, 5, 10}, {5, 6, 10}, {6, 10, 12}	{2, 3}, {2, 4}, {2, 3, 4}, {2, 3, 9}, {2, 3, 11}, {2, 3, 13}, {2, 4, 9}, {2, 4, 11}, {2, 4, 13}, {3, 4, 9}, {3, 4, 11}, {3, 4, 13}, {2, 3, 4, 9}, {2, 3, 4, 11}, {2, 3, 4, 13}, {2, 3, 9, 11}, {2, 3, 9, 13}, {2, 3, 11, 13}, {2, 4, 9, 11}, {2, 4, 9, 13}, {2, 4, 11, 13}, {3, 4, 9, 11}, {3, 4, 9, 13}, {3, 4, 11, 13}, {2, 3, 9, 11, 13}, {2, 4, 9, 11, 13}, {3, 4, 9, 11, 13}
2	{5, 7}, {5, 10}, {6, 7}, {7, 10}, {7, 12}, {3, 10, 12}, {6, 10, 12}	{5, 7}, {5, 10}, {7, 10}, {4, 5, 7}, {4, 5, 10), {4, 7, 10}, {5, 6, 7}, {5, 6, 10}, {6, 7, 10}, {4, 5, 6, 7}, {4, 5, 6, 10}, {4, 6, 7, 10}
3	{2, 6}, {2, 7}, {5, 7}, {6, 7}, {7, 10}, {7, 12}, {5, 6, 10}, {6, 10, 12}	{2, 6}, {2, 7}, {6, 7}
4	{7}	{7}, {1, 7}, {7, 8}. {7, 11}, {1, 7, 8}, {1, 7, 11}, {7, 8, 11}, {1, 7, 8, 11}
5	{3, 12}, {6, 12}, {6, 7}, {7, 12}	{3, 12}, {6, 12}, {7, 12}, {3, 6, 7}, {3, 6, 12}, {3, 7, 12}, {6, 7, 12}. {3, 6, 7, 12}

4 Comparison of Karos and Peters as Well as Mercik and Lobos Approaches in Example of Corporate Network

Both approaches (the Karos and Peters [13] and Mercik and Lobos [14]) provide indices that measure the control power of all firms involved in corporate networks. Here, we briefly present these methods and then apply these approaches to a theoretical example of the shareholding structure presented in the previous section.

4.1 Karos and Peters Approach

In 2015, Karos and Peters [13] introduced an approach to modeling and measuring the indirect control power of firms in a shareholding structure. They model indirect control in two equivalent ways: by the invariant mutual control structure (a map that assigns a set of controlled players to each coalition), and by a simple game structure (i.e., a vector of simple games where each simple game indicates who controls the corresponding player). For each firm in a corporate shareholding network, there is a (monotonic) simple game whose winning coalitions are exactly those that control that company. Karos and Peters proposed the Φ index to measure indirect control:

$$\Phi_i(C) = \sum_{k \in N} \sigma_i(v_k^C) - v_i^C(N), \tag{1}$$

where σ is the Shapley-Shubik index, and v_k^C is a simple game indicates who directly or indirectly controls Player k for invariant mutual structure C.

Let us calculate Φ in the theoretical example, taking into account the information from Table 2 and (1). For each company, Table 2 provides the set of minimal winning coalitions. Hence, we have uniquely defined a simple game for each company, and we can calculate the Shapley-Shubik index. Then, using Formula (1) of the Φ index, we obtain the control power calculated for each firm. The results are presented in Table 3. Note that each company is controlled by N (i.e., all firms in the theoretical example), and each investor is controlled by no firm by definition. Thus, following the Karos and Peters definitions of the index, we have $v_i(N) = 1$ for each company $i = 1, 2, 3, 4, 5$, and $v_i(S) = 0$ for each investor $i = 6, 7, 8, 9, 10, 11, 12, 13$ and $S \subseteq N$.

The last column in Table 3 gives the Φ index for all firms; thus, we can see, that it ranks the companies from most- to least-powerful as follows: 2, 5, 3, 4, 1. For the investors, the ranking is as follows: 7, 6, 12, 10, 9 = 11 = 13, 8. So, the most-powerful in the sense of power control is Investor 7, whereas Investor 8 (being a dummy player) is least-powerful.

4.2 Mercik and Lobos Approach

In [14], Mercik and Lobos proposed a measure of reciprocal ownership called the index of implicit power as a modification of the Johnston power index. Mercik and Lobos suggested a three-step algorithm to calculate the implicit power index. In Step 1, the absolute value of the Johnston index is calculated for each company, taking only the direct ownerships into account. Then, in Step 2, each value of the power index

Table 3. Calculations of Φ index in theoretical example.

Firm	Power distribution in accordance with σ index in simple game v_i							
	$i = 1$	$i = 2$	$i = 3$	$i = 4$	$i = 5$	$i = 6, \ldots, 13$	Total	Φ
1	0.000	0.000	0.000	0.000	0.000	0.000	0.000	-1.000
2	0.196	0.000	0.133	0.000	0.000	0.000	0.329	-0.671
3	0.121	0.017	0.000	0.000	0.083	0.000	0.221	-0.779
4	0.098	0.000	0.000	0.000	0.000	0.000	0.098	-0.902
5	0.056	0.150	0.050	0.000	0.000	0.000	0.256	-0.744
6	0.096	0.067	0.250	0.000	0.250	0.000	0.662	0.662
7	0.265	0.400	0.433	1.000	0.250	0.000	2.348	2.348
8	0.000	0.000	0.000	0.000	0.000	0.000	0.000	0.000
9	0.010	0.000	0.000	0.000	0.000	0.000	0.010	0.010
10	0.092	0.267	0.083	0.000	0.000	0.000	0.442	0.442
11	0.010	0.000	0.000	0.000	0.000	0.000	0.010	0.010
12	0.047	0.100	0.050	0.000	0.417	0.000	0.613	0.613
13	0.010	0.000	0.000	0.000	0.000	0.000	0.010	0.010

calculated in Step 1 is divided equally for each shareholder-company among all of its shareholders. They call this a first-degree regression. In Step 3, the absolute value of the implicit power index is calculated for each company by summing up the appropriate values in the whole corporate network. For each investor, the absolute value of the implicit power index is calculated by summing up the appropriate values across the entire system of companies. Then, these absolute values are appropriately standardized to obtain the implicit power index of each firm.

Now, let us calculate the implicit power index in the theoretical example (see results in Tables 4 and 5). Table 4 shows the results of the realization of Step 1 of the Mercik and Lobos procedure; i.e., the distribution of the absolute Johnston index in each company. To find the distribution of the γ index, it is necessary to know all of the vulnerable coalitions for each company. The sets of corresponding vulnerable coalitions are given in Table 2.

Table 5 presents the results obtain after the realization of Steps 2 and 3 of the Mercik and Lobos procedure; i.e., the absolute and standardized values of the implicit power index in the theoretical example.

Looking at the result of the calculations in Table 5, we obtain the following rankings of the investors and companies according to the implicit index (from most- to least-powerful firm). Specifically, the ranking of investors is 7, 6, 10, 12, 11, 8, 9 = 13, while the ranking of companies is 1, 2, 4 = 5, 3.

4.3   Comparison of Methods

In this section, we compare two power indices introduced by Karos and Peters [13] and Mercik and Lobos [14], respectively. These indices were just submitted to the comparison in [15], where two real-world examples of corporate shareholding structures (one cyclic and the other acyclic) were considered. Here, we focus on the ranking of

Table 4. Distribution of absolute γ index in Step 1 of Mercik and Lobos algorithm.

Firm	Company				
	1	2	3	4	5
1 and 8					
2	12.0000		1.0000		
3	7.0000				0.8333
4	7.0000				
5		4.0000			
6			1.0000		1.3333
7		4.0000	1.0000	8.0000	0.8333
9	0.3333				
10		4.0000			
11	0.3333				
12					5.0000
13	0.3333				

Table 5. Absolute and standardized values of implicit power index in example.

Investor	Company					Absolute index of investor	Standardized index of investor
	1	2	3	4	5		
6	4.7333	1.0000	1.2000		1.7500	8.6833	0.1820
7	6.4833	5.0000	1.2000	8.0000	1.2500	21.9333	0.4597
8	1.7500					1.7500	0.0367
9	0.3333					0.3333	0.0070
10	2.4000	4.0000	0.2000			6.6000	0.1383
11	2.0833					2.0833	0.0437
12		1.0000			5.0000	6.0000	0.1257
13	0.3333					0.3333	0.0070
Absolute index of company	18.1167	11.0000	2.6000	8.0000	8.0000	47.7167	
Standardized index of company	0.3797	0.2305	0.0545	0.1677	0.1677		

firms provided by these two approaches, and we also consider the more-complex example of a corporate network; i.e., that which is presented in Sect. 3.

Every power index gives a total ranking of players so that the control power for any two arbitrary firms is always comparable. Some approaches to indirect control only assess the control power for investors (like the Crama and Leruth method [10], for example). The Karos and Peters [13] and Mercik and Lobos [14] approaches give the assessment of control power for all firms involved in the corporate network. If we intend to measure control power like influence, we are interested in rankings rather than the "exact" values of the power indices.

In [15] the control power of firms in two real-world examples of corporate share-holding structures were calculated, and these two approaches resulted in the same rankings of investors and companies. The Φ index proposed by Karos and Peters [13] can be seen as a modification of the Shapley-Shubik index [18], whereas the implicit index is a modification of the Johnston index [20]. Also, the Shapley-Shubik, Banzhaf, and Johnston indices are qualitatively equivalent in semi-complete games (see [30]). The class of semi-complete games is larger than the class of weighted majority games. Thus, we suspected that the indices used by these approaches will also rank players in the same way. However, our suspicion turned out to be unjustified. More precisely, we found an example of a shareholding structure in which the rankings of companies and investors provided by these two methods differ. Namely, the results obtain in Sects. 4.1 and 4.2 for the two indices and the theoretical example show that the Φ and implicit indices are not qualitatively equivalent; i.e., their give different rankings for the firms. According to Φ, the least-powerful investor is Firm 8, which is a null player (as Firm 8 does not belong to any minimal winning coalition). For the implicit index, the least-powerful are Investors 9 and 13; they have the same control power, which is lower than the control power of Investor 8 (see Table 5). According the implicit index, Investors 9 and 13 have the least control power, even though each of these investors is critical for two minimal winning coalitions. More precisely, Investor 9 is critical in coalitions $\{3, 4, 9\}$ and $\{3, 7, 9\}$, whereas Investor 13 is critical in $\{3, 4, 13\}$ and $\{3, 7, 13\}$ (see Table 2). The coalition composed of Companies 3 and 4 cannot control Company 1 by themselves. However, if they convince Investor 9 or 13 to joint them, then these three firms can jointly directly control Company 1. Similarly, Company 3 and Investor 7 together do not have enough voting rights to control Company 1 on the one hand. On the other hand, Coalitions $\{3, 7, 9\}$ and $\{3, 7, 13\}$ indirectly control Company 1. Why does it happen that Investor 8 has greater control power than either Investors 9 or 13? This is because, in Step 2 of the Mercik and Lobos algorithm, the value assigned to a company in Step 1 is equally divided among all of its shareholders. Hence, Company 4, being critical in 15 vulnerable coalitions that control Company 1 (see Table 2), obtains more power according to the absolute Johnston index in Step 1 than Investors 9 or 13. Then, the value assigned to Company 4 is equally shared among its four shareholders. Investor 8 is one of the shareholders of Company 4. This value assigned to Investor 8 is greater than the value assigned to Investors 9 and 13 in Step 1. Precisely, both Investors 9 and 13 obtain 0.33 each in our example; also, Company 4 receives 7, and as con-sequence, Investor 8 obtains $0.25 \cdot 7 = 1.75$ (see Tables 4 and 5). If the implicit index did not equally share the slices obtained in Step 1, then Investor 8 would receive nothing according to the absolute Johnston index (for example). This would happen because Investor 8 is not critical in any vulnerable coalition.

The indices considered in this paper are not qualitatively equivalent, as we can see in the theoretical example. There is, of course, not only one difference between these indices. Taking into account the companies and the values assigned to them, we see that the rankings are totally different. For the Φ index, the most-powerful company is Firm 2, and the least-powerful is Company 1. The control power of Company 3 is in the middle. For the implicit index, the most-powerful is Company 1, and the least-powerful is Company 3. Company 2 is in second place in the order from most- to least-powerful companies. The reason is that these indices measure different aspects of the position of the companies. We can say that the Φ index measures the companies' power control in other companies, but the implicit index rather measures the position of a company in the whole network. As we can see in the considered example, according to the implicit index, Company 1 is more powerful among the stock companies in the network and has the greatest number of shareholders; hence, it has the greatest number of connections with other firms in the shareholding structure. Therefore, the implicit index could measure the "thickness" of the network around the companies. The more investors, the more power that is assigned. Maybe company power means the power of the "middleman" position in the corporate network according to the implicit index. If we remove a middleman from a given network, we completely disconnect the connections between the nodes that it brokers. Thus, the implicit index determines the potential breakup power of a given network for a given stock company. This is our hypothesis. In the theoretical example, removing Company 1 from the network means to remove seven arcs and three nodes (including the node that refers to Company 1). Removal of any other company in Fig. 1 does not destroy such a considered network. In accordance with the implicit index, the second-most-powerful company is Company 2. Eliminating Company 2 means also eliminating seven arcs but only two nodes from the network. Eliminating Company 4 provokes the disappearance of six direct linkages and two nodes. Company 4 together with Company 5 is in the third position in the ranking of the implicit index. Removing Company 5 implies the elimination of five direct connections and two nodes. Removal of Company 3 implies the elimination of five direct connections and one node. So, in our case for stock companies, the rankings given by the implicit index are almost the same as those given by the breakup power.

Do more direct shareholders implicate more power? Does the greater number of direct shareholders imply the greater value of these indices? Taking into account the implicit index, the answer to both questions is "yes." If a company has a lot of investors, it has a greater number of vulnerable coalitions as a consequence. This, in turn, implies the greater value of the absolute Johnston index and greater value assigned to the origin company as a consequence. On the other hand, if a company has stock companies as shareholders who, in turn, have investors who are shareholders in the original (first) company, it also implies an increase in the value assigned to the origin company. In our example, 27 vulnerable coalitions directly control Company 1. Company 4 is a shareholder of Company 1. Investor 11 is a shareholder of both Companies 1 and 4 (see Table 1 and Fig. 1). The sum of the absolute Johnston index over all shareholders of Company 1 is equal to 27. From these 27, 7 are assigned to Company 4 and 1/3 to Investor 11. Then, the value 7 assigned to Company 4 in Step 2 is subsequently equally divided among the shareholdings of Company 4, which results in the fact that Investor 11 obtains an additional $7/4 = 1.75$ in Step 2. And this last value (1.75) increases, the value is assigned to Company 1 in Step 3 (see Table 5).

Regarding the Karos-Peters approach, more control power in other companies results in greater power. This can be deduced from the formula of the Φ index as well as from Table 3. In Table 3, we see that, for particular firm i, its power is calculated according to the Shapley-Shubik index over all other firms; then, the values obtained are summed.

What are the other factors that influence the power of companies? This requires more in-depth research on these indices.

5   On Reasonable Requirements for Indirect Control Measure

In this section, we pose the question of the set of required properties for a good measure of indirect control in the corporate shareholding structures. Let the following text taken from [31, p. 676] be our motto for this section:

> "The axiomatic characterization of power indices is a main topic in the field for at least two reasons. First, characterizing a rule by means of a set of properties may be more appealing than just giving its explicit definition. Second, deciding on whether to use a rule or another in a particular situation may be done more easily taking into account the properties that each rule satisfies. In fact, many power indices have shown to have different sets of properties that determine them uniquely."

In [13], Karos and Peters started from the set of axioms and arrived with one index (Φ). More precisely, they considered the following properties: the anonymity, null player, transfer, constant-sum, and control player properties. The *constant-sum postulate* normalizes the power index over different mutual control structures and acts the role of the known efficiency property. The null player and anonymity axioms are practically the same as the known properties with the same names in the literature. The anonymity also refers to the symmetry axiom. *The controlled player condition* states that, if stock Company j is controlled by at least one coalition (and thus by all firms) but does not exercise any control, then the power of Company j is settled at -1. Furthermore, if i is an investor (that is to say, she/he is controlled by no coalition at all) and Firms i and j exert the same marginal control with respect to any coalition, then their difference in power is settled at 1; i.e., Investor i obtains 1 more than Company j. The *transfer property* states that, if going from a mutual structure C' to C involves exactly the same increase in control power as going from mutual structure D' to D, then the power of each firm should change by the same quantity when going from C' to C as when going from D' to D. The transfer property is related to a property with the same name proposed by Dubey [32] in order to be axiomatized the Shapley-Shubik index.

The question about the inescapable properties in the indirect control aspect is interesting. Let us look at some properties and try to formulate some in the context of indirect control.

Symmetry Property. The symmetry property is common to all power indices. In the indirect control context, the symmetry postulate seems desirable, and the indices considered here satisfy this property. The justification of this property is that the firms in the same position in the corporate network should have the same control power. The same position means that the firms have the same voting rights in the same companies and are controlled by the same firms (if any).

Null Player Property. Regarding null player property, we can say that it is also rather common to all power indices. The Φ index satisfies this postulate, and the Johnston index also fulfills it; however, the implicit index does not. What does this property mean in the corporate networks? A firm's control power is null unless its voting rights can transform at least one losing coalition into a winning one. Why does the implicit index not satisfy the null player property? The Johnston index satisfies this property; however, in Stage 2 of the procedure for calculating the implicit index, the value assigned to each company in Stage 1 is equally divided among all of the shareholders of the company. In this way, dummy players can obtain something, which is exactly what happened for Investor 8 in the theoretical example considered in this paper (see Table 5). One way to make the implicit index satisfy the null player property is to change Step 2 in the Mercik and Lobos algorithm. Namely, we can divide the quantity assigned to a company in Stage 1 among only the critical players in vulnerable coalitions according the proportion indicated by the absolute Johnston index.

Efficiency Property. The efficiency postulate is always problematic in the context of power indices. That is to say, it depends on whether we define the power calculated by the power indices as the power to share a purse or as the power to influence the outcome. In the former case, the efficient property is required, and in the latter case, it is not. By definition, the implicit index satisfies the efficiency property separately for investors and companies. This means that the sum of the implicit index over all investors is equal to 1, and the sum of the implicit index over all companies equals 1 as well. The Φ index does not fulfill the null player axiom as it was defined in Sect. 2. However, the sum of Φ over all firms is always equal to zero (and as mentioned above, Φ satisfies the constant-sum axiom).

Non-negativity Property. If we would like to have a measure of indirect power control that assigns to the firms' non-negative values or if we would like to have a relative measure, then this property is indispensable. The Φ index does not satisfy this property; however, the values assigned to the investors by this index is always non-negative by definition. The implicit index as the relative measure satisfies this condition.

Below, we propose some formulations of the bloc, dominance, and donation properties in an indirect control context. These properties seem to be reasonable requirements for indirect control measure.

Bloc Property. If two firms merge, then the new firm should not have less control power than each of the previous firms.

Dominance Property. Let us consider a corporate network in which two firms are almost in the same position. This means they are controlled by the same firms with the same amount of voting rights (if any) and have the same number of voting rights in the same companies except one company, where one of the considered firms has more voting rights than other. Then a firm with a larger block of voting rights cannot receive less control power in the whole network than a firm with a smaller block of voting rights.

Donation Property. In a corporate structure, a firm that donates its voting rights to another firm should not increase its control power.

Null Player Removable Property. The null player removable property states that after removing the null firms (i.e., the firms whose voting rights cannot transform any losing coalition into a winning one), the non-null firms' measures of control power should remain unchanged.

What other properties are desirable for a good measure of indirect control? It remains an open problem and a topic for a future paper.

6 Conclusion and Further Developments

The issue of indirect control is complex and important in corporate governance and financial economics. In this paper, we gave a short literature review of the most-recently-introduced game-theoretical approaches to assess the voting power of firms involved in complex corporate shareholding networks. Then, we analyzed two approaches (those of Karos and Peters [13] and Mercik and Lobos [14]), and we focused on the rankings of the firms that these two methods provide (see Sects. 4.1 and 4.2). We found an example where these rankings are different not only for the investors (firms without shareholdings) but also for the companies. This fact reflects that these methods highlight different aspects of indirect control. Since, in particular, the rankings for the companies in our example differ quite widely, we tried explain the meaning of the values assigned to the companies by the power indices (see Sect. 4.3). We also presented a short discussion on the reasonable requirements for indirect control measure (see Sect. 5).

One of further development can be the continuation of comparing the game-theoretical methods that measure shareholding indirect control, not only for the investors (ultimate shareholders) but also for the stock companies involved in corporate shareholding networks. To the authors' knowledge, there are only three methods of this kind. Specifically, those of Karos and Peters [13], Mercik and Lobos [14], and Levy and Szafarz [3].

Exactly which properties should be met by a good power index in the context of indirect control still remains an interesting open problem.

Modification of the implicit index of Mercik and Lobos [14] and changing Step 2 of its algorithm in particular can be an idea to satisfy the null player property by this index.

Indirect control in complex corporate shareholding networks is realized by direct and indirect linkages. Therefore, further developments may take into account the possible application of the idea of Forlicz, Mercik, Ramsey, and Stach [33] proposed for multigraphs and, hence, evaluate the importance of not only the firms but also the importance of mutual connections in the corporate networks.

Last but not least, an open problem is the lack of computer programs to make these calculations. It is really problematic to apply game-theoretical approaches without efficient algorithms.

Acknowledgements. Research is financed by the statutory funds (no. 11/11.200.322) of the AGH University of Science and Technology and by statutory funds of WSB University in Wroclaw. The authors would also like to thank the comments and suggestions of Gianfranco Gambarelli and Cesarino Bertini. Moreover, authors thank the anonymous reviewers for their careful reading of the manuscript and helpful comments and suggestions.

References

1. Claessens, S., Djankov, S., Lang, L.H.P.: The separation of ownership and control in East Asian corporations. J. Financ. Econ. **58**, 81–112 (2000)
2. Faccio, M., Lang, L.H.P.: The ultimate ownership of Western European corporations. J. Financ. Econ. **65**, 365–395 (2002)
3. Levy, M., Szafarz, A.: Cross-ownership: a device for management entrenchment? Rev. Financ. **21**(4), 1675–1699 (2017)
4. Gambarelli, G., Owen, G.: Indirect control of corporations. Int. J. Game Theory **23**, 287–302 (1994)
5. Turnovec, F.: Privatization, ownership structure and transparency: how to measure the true involvement of the state. Eur. J. Polit. Econ. **15**, 605–618 (1999)
6. Denti, E., Prati, N.: An algorithm for winning coalitions in indirect control of corporations. Decis. Econ. Financ. **24**, 153–158 (2001)
7. Denti, E., Prati, N.: Relevance of winning coalitions in indirect control of corporations. Theory Decis. **56**, 183–192 (2004)
8. Hu, X., Shapley, L.S.: On authority distributions in organizations: controls. Games Econ. Behav. **45**, 153–170 (2003)
9. Hu, X., Shapley, L.S.: On authority distributions in organizations: equilibrium. Games Econ. Behav. **45**, 132–152 (2003)
10. Crama, Y., Leruth, L.: Control and voting power in corporate networks: concepts and computational aspects. Eur. J. Oper. Res. **178**, 879–893 (2007)
11. Levy, M.: Control in pyramidal structures. Corp. Gov. Int. Rev. **17**, 77–89 (2009)
12. Levy, M.: The Banzhaf index in complete and incomplete shareholding structures: a new algorithm. Eur. J. Oper. Res. **215**, 411–421 (2011)
13. Karos, D., Peters, H.: Indirect control and power in mutual control structures. Games Econ. Behav. **92**, 150–165 (2015)
14. Mercik, J., Łobos, K.: Index of implicit power as a measure of reciprocal ownership. In: Nguyen, N.T., Kowalczyk, R., Mercik, J. (eds.) Transactions on Computational Collective Intelligence XXIII. LNCS, vol. 9760, pp. 128–140. Springer, Heidelberg (2016). https://doi.org/10.1007/978-3-662-52886-0_8
15. Bertini, C., Mercik, J., Stach, I.: Indirect control and power. Oper. Res. Decis. **26**(2), 7–30 (2016)
16. Kołodziej, M., Stach I.: Control sharing analysis and simulation. In: Sawik, T. (ed.) Conference Proceedings of ICIL 2016: 13th International Conference on Industrial Logistics, Zakopane, Poland, pp. 101–108. AGH University of Science and Technology, Krakow (2016)
17. Stach, I.: Indirect control of corporations: analysis and simulations. Decis. Making Manuf. Serv. **11**(1–2), 31–51 (2017)
18. Shapley, L.S., Shubik, M.: A method for evaluating the distributions of power in a committee system. Am. Polit. Sci. Rev. **48**, 787–792 (1954)
19. Banzhaf, J.F.: Weighted voting doesn't work. A mathematical analysis. Rutgers Law Rev. **19**, 317–343 (1965)
20. Johnston, R.J.: On the measurement of power: Some reactions to Laver. Environ. Plan. A **10**, 907–914 (1978)
21. Holler, M.J.: Forming coalitions and measuring voting power. Polit. Stud. **30**, 262–271 (1982)
22. Crama, Y., Leruth, L.: Power indices and the measurement of control in corporate structures. Int. Game Theory Rev. **15**(3), 1–15 (2013)

23. Stach, I.: Shapley-Shubik index. In: Dowding, K. (ed.) Encyclopedia of Power, pp. 603–606. SAGE Publications, Los Angeles (2011)
24. Bertini, C., Stach, I.: Banzhaf voting power measure. In: Dowding, K. (ed.) Encyclopedia of Power, pp. 54–55. SAGE Publications, Los Angeles (2011)
25. Bertini, C., Freixas, J., Gambarelli, G., Stach, I.: Comparing power indices. Int. Game Theory Rev. **15**(2), 1340004-1–1340004-19 (2013)
26. Bertini, C., Stach, I.: On public values and power indices. Decis. Making Manuf. Serv. **9**(1), 9–25 (2015)
27. Stach, I.: Power measures and public goods. In: Nguyen, N.T., Kowalczyk, R., Mercik, J. (eds.) Transactions on Computational Collective Intelligence XXIII. LNCS, vol. 9760, pp. 99–110. Springer, Heidelberg (2016). https://doi.org/10.1007/978-3-662-52886-0_6
28. Felsenthal, D.S., Machover, M.: The Measurement of Voting Power: Theory and Practice, Problems and Paradoxes. Edward Elgar Publishers, London (1998)
29. Laruelle, A.: On the choice of a power index. Working Papers AD 1999-20, Instituto Valenciano de Investigaciones Económicas (Ivie) (1999)
30. Freixas, J., Marciniak, D., Pons, M.: On the ordinal equivalence of the Johnston, Banzhaf and Shapley power indices. Eur. J. Oper. Res. **216**(2), 367–375 (2012)
31. Álvarez-Mozos, M., Ferreira, F., Alonso-Meijide, J.M., Pinto, A.A.: Characterizations of power indices based on null player free winning coalitions. Optim. J. Math. Program. Oper. Res. **64**(3), 675–686 (2015)
32. Dubey, P.: On the uniqueness of the Shapley value. Am. Polit. Sci. Rev. **4**, 131–139 (1975)
33. Forlicz, S., Mercik, J., Ramsey, D., Stach, I.: The Shapley value for multigraphs. In: Nguyen, N., Pimenidis, E., Khan, Z., Trawiński, B. (eds.) Computational Collective Intelligence. LNAI, vol. 11056, 10th International Conference, ICCCI 2018 Bristol, UK, September 5–7, 2018 Proceedings, Part II, pp. 213–221. Springer, Cham (2018). https://doi.org/10.1007/978-3-319-98446-9_20

The Effect of Brexit on the Balance of Power in the European Union Council Revisited: A Fuzzy Multicriteria Attempt

Barbara Gładysz[1](✉), Jacek Mercik[2], and David M. Ramsey[1]

[1] Faculty of Computer Science and Management,
Wrocław University of Science and Technology, Wrocław, Poland
{barbara.gladysz, david.ramsey}@pwr.edu.pl
[2] WSB University in Wrocław, Wrocław, Poland
jacek.mercik@wsb.wroclaw.pl

Abstract. The approaching exit of Great Britain from the European Union raises many questions about the changing relations between other member states. In this work, we propose a new fuzzy game for multicriteria voting. We use this game to show changes in Shapley's values in a situation where the weights of individual member countries are not determined and we describe non-determinism with fuzzy sets. In particular, this concerns considerations related to pre-coalitions.

Keywords: Voting games · Power index · Fuzzy set
European Union Council

1 Introduction

Mercik and Ramsey (2017) presents the analysis of the change in power relations in the European Union Council. In this approach, it was assumed that possible pre-coalitions of member countries would arise based on the similarities and gathering of countries around the leading (i.e. the biggest) countries of the European Union. These six states (Germany, France, the United Kingdom, Italy, Spain and Poland) are significantly larger (according to population) than the others. This approach results from the following fact: in order to be able to block any decision of the Council of Ministers of the Union, at least 4 countries representing at least 35% of the Union's population are needed. In such cases, each such coalition must contain 3 of these large countries, otherwise being ineffective.

At the end of their paper, Mercik and Ramsey (2017) write "since the UK is a somewhat isolated (or independent) player in the EU council (both in practice and in accordance with our model). Brexit is unlikely to lead to any major rearrangement of natural coalitions within the EU. On the other hand, the fact that such a large country is leaving the EU might have a significant impact on the balance of power within the EU council. According to the model presented here, the pre-coalition formed around Poland will lose power, since France (or Germany) can now form a blocking coalition

© Springer-Verlag GmbH Germany, part of Springer Nature 2018
N. T. Nguyen et al. (Eds.): TCCI XXXI, LNCS 11290, pp. 80–91, 2018.
https://doi.org/10.1007/978-3-662-58464-4_8

with any other pre-coalition. However, since France and Germany have tended to co-operate together, it is likely that Poland's power in the pre-Brexit game is exaggerated."

The basic problem in the model with classic power index (Shapley value) is that all possible coalitions (pre-coalitions) are treated as determined coalitions, i.e. their weights are a simple sum of the weights of individual countries. This corresponds to the concept of party discipline in contemporary parliaments, where it is also assumed that members of a given grouping (party) vote in the same way as its leader. Simple parliamentary observations show this is, in most cases, a false assumption. Parliamentarians for various reasons deviate from party discipline and therefore the classic Shapley-Shubik power index (1954) needs to be modified. What's more, we believe that with regard to countries, the expectation that they will follow the leading countries without looking at their national interests is all the more untrue. In this article, we propose a fuzzy approach to the problem of analyzing changes in the balance of power between the member states of the European Union.

The article is structured as follows: after the introductory chapters on the basics of simple games and fuzzy sets, we present the proposed fuzzy definition of the Shapley-Shubik power index. We then show its application to the European Union Council. We attempt to provide the differences between the classic and fuzzy Shapley-Shubik index, and their possible explanations. The article ends with conclusions and plans for future research.

2 Initial Assumptions

Simple games (voting games) are a subgroup of cooperative games in which the value of a given coalition depends on the result of voting in such a way that if the result of voting is consistent with the vote of a given coalition, its characteristic function equals 1.

Let's assume that we are dealing with $N \geq 3$ participants (players) of a given vote. The characteristic function specified on $T \subseteq N$ is a non-negative real function v on the set of all coalitions of players N, $v: T \to v(T)$ where $T \subseteq N$, $v(T) \geq 0$. Evidently, $v(\emptyset) = 0$ since the coalition of all players (the so called grand coalition) cannot gain more than what is gained when all the players cooperate together.

If a characteristic function, $v(T)$, accepts only 0 or 1 values, we are dealing with a simple strict majority game.

$$v(T) = \begin{cases} 1 & if \sum_{i \in T} w_i \geq q \\ 0 & otherwise \end{cases} \tag{1}$$

when $q > \frac{\sum_{i \in T} w_i}{2}$.

In anticipation of further results, it must stipulate here that, although we will continue to deal with simple strict majority game, the introduction of fuzziness will disrupt this simple game. For certainty, each such game will be referred to as fuzzy simple strict majority game.

A cooperative game $(N,)$ is called superadditive if for each pair of coalitions S, T we have $(S \cup T) \geq v(S) + v(T)$ when $S \cap T = \emptyset$. This condition ensures that it pays to cooperate[1].

Any vector $(x_1, \ldots, x_N) \in R^N$ such that $x_1 + \ldots + x_N = v(N)$ is a payoff vector of the game $(N,)$. The set of all possible payoff vectors of the game $(N,)$ given by (v).

The set of payoff vectors (v) is called a solution of the cooperative game defined by the characteristic function v on the set of players N when (v) is a subset of $P(v)$ and this set of payoff vectors satisfies the concept of a cooperative solution accepted by the players. For example, they might only accept egalitarian solutions (where each player obtains the same payoff) or solutions where each player's payoff is proportional to their contribution (this rule is commonly used to split the profits from a joint investment), etc.

A set (v) is called a value if for any game $(N,)$ the set $r(v)$ contains exactly one member, i.e. there only exists one solution of a game. Shapley shows (so called Shapley's theorem, 1953) that the only effective, additive and symmetric value of a game satisfying the null player condition is the Shapley value, $\varphi(v)$, for the cooperative game (N, v): $\varphi(v) = (\varphi_1(v), \varphi_2(v), \ldots, \varphi_N(v))$,

$$\varphi_i(v) = \sum_{T \subseteq N, i \in T} \frac{(t-1)!(n-t)!}{n!} (v(T) - v(T \setminus \{i\})) \tag{2}$$

This is the expected marginal value of a player to a coalition under the assumption that each player enters the coalition in a random order, i.e. all coalitions are equally likely, until the grand coalition is formed. The normalized Shapley value, resulting from the use of 0–1 characteristic function, is called the Shapley-Shubik power index (1954). The Shapley value has a number of properties that give it possible interpretations and applications in many areas. These properties include: effectiveness, a null player property, symmetry (also called anonymity), local monotonicity, and strong monotonicity[2]. In particular, when determining fuzzy games, we will use a property called additivity (also called the "law of aggregation"). This property states that when two independent games are combined, their values must be added player by player.

3   Fuzzy Cooperative Games

The actual use of simple games to analyze group decision situations (undertaken by more than two players) should be naturally made fuzzy. Although it commonly happens, the decision-makers should not be described in a determined way, because in the majority of votes observed, their number and behavior are not constant. Although

---

[1] It is worth noting that in Gładysz and Mercik (2018) the condition of superadditiveness is not met for simple fuzzy games. The condition can be modified for the so-called partial superadditiveness.

[2] There are several theorems related to additional properties of the Shapley value, e.g. Young (1985) proved that the Shapley value is the only value satisfying the properties of effectiveness, symmetry and strong monotonicity; van den Brink (2001) showed that it is the only value preserving a fairness condition according to a modification of marginal contributions to a coalition; Myerson (1977) showed that it preserves fairness based on balanced contributions to a coalition.

various security mechanisms are introduced (such as party discipline in parliamentary elections, appeals for national voting in international bodies, etc.), however, real votes are almost always different (colloquially understood as fuzzy). You can therefore fuzzy the basic elements of cooperative games:

- Borkotokey (2008) and Yu and Zhang (2010) introduced fuzzy coalition and fuzzy characteristic function,
- Tsurumi et al. (2001), Li and Zhang (2019) used the Choquet integral to cooperative game with real characteristic function,
- Pang et al. (2014) used concave integral, and
- Meng and Zhang (2011) used the support of fuzzy number.

In the analysis of the European Union Council, which is a continuation of the research presented in Mercik and Ramsey (2017), we propose the use of fuzzy cooperative game with real characteristic function and fuzzy weights. For this purpose, following Gladysz and Mercik (2018), we apply satisfaction measure. In this work, we suggest a blurred multi-criteria game in which players' weights are fuzzy sets and the characteristic function of a coalition is a real function defined on a set of fuzzy numbers. Similarly to Gladysz and Mercik (2018), we use the measure possibility to construct a characteristic function. Thus, the characteristic function determines the possibility of a given coalition to win.

We will now present the basic concepts from the theory of fuzzy sets. Zadeh (1965) introduced the concept of a fuzzy set. Fuzzy set $\widetilde{A}$ in space X is a set of ordered pairs: $\{(x, \mu_A(x) : x \in X)\}$ where $\mu_A : X \to [0, 1]$ is a function of belonging to a fuzzy set. We will represent the fuzzy set as $\widetilde{A} = x_1|\mu_1 + \cdots + x_n|\mu_n$. For example, a fuzzy number $\widetilde{A} = 1|0.5 + 2|1 + 3|0$ we interpret as: most possibly 2, but it is ½ possible that the number will take the value 1, the value 3 is impossible.

If we have two fuzzy sets $\widetilde{A}$ and $\widetilde{B}$, the degree of belonging of a given element x to a set $\widetilde{A} \cap \widetilde{B}$ is equal to:

$$\mu_{A \cap B}(x) = min\{\mu_A(x), \mu_B(x)\}. \tag{3}$$

Dubois and Prade (1978) introduced the concept of interval fuzzy numbers $\widetilde{X}$ family of real closed intervals $[\widetilde{X}]_\lambda$ where $\lambda \in [0, 1]$ such that: $\lambda_1 < \lambda_2 \Rightarrow [\widetilde{X}]_{\lambda_1} \subset [\widetilde{X}]_{\lambda_2}$ and $I \subseteq [0, 1] \Rightarrow [\widetilde{X}]_{supI} = \bigcap_{\lambda \in I} [\widetilde{X}]_\lambda$. The range $[\widetilde{X}]_\lambda$ for a determined $\lambda \in [0, 1]$ is called λ-level of fuzzy number $\widetilde{X}$. We will mark it as $[\widetilde{X}]_\lambda = [\underline{x}(\lambda), \bar{x}(\lambda)]$.

The fuzzy number $\widetilde{X}$ is called a LR fuzzy number if its membership function takes the form (Dubois and Prade, 1978):

$$\mu_X(x) = \begin{cases} L\left(\frac{m-x}{\alpha}\right) & for & x < \underline{m} \\ 1 & for & \underline{m} \le x \le \bar{m} \\ R\left(\frac{x-\bar{m}}{\beta}\right) & for & x > \bar{m} \end{cases} \tag{4}$$

where: $L(x), R(x)$ – continuous nonincreasing functions x; $m, \alpha, \beta > 0$

Functions $L(x)$, $R(x)$ are called the shape functions of a fuzzy number. The most commonly used shape functions are: $\max\{0, 1 - x^p\}$ and $\exp(-x^p)$, $x \in [0, +\infty)$, $p \geq 1$. A range fuzzy number for which $L(x). R(x) = \max\{0, 1 - x^p\}$ and $\underline{m} = \bar{m} = m$ is called a triangular fuzzy number, it will be represented as (m, α, β).

Let $\tilde{X}, \tilde{Y}$ be two fuzzy numbers with membership functions, respectively $\mu_X(x), \mu_Y(y)$ and let $z = f(x, y)$. Then, according to the extension principle of Zadeh (Zadeh 1965), the function of belonging to the fuzzy set $\tilde{Z} = f(\tilde{X}, \tilde{Y})$ takes the form:

$$\mu_Z(z) = sup_{z=f(x,y)}(min(\mu_X(x), \mu_Y(y))) \tag{5}$$

If we want to compare two fuzzy sets, that is, to determine the possibility that realization $\tilde{X}$ is greater/equal (not less) than the realization $\tilde{Y}$, then we can use the index proposed by Dubois and Prade (1988):

$$Pos(X \geq Y) = sup_{x \geq y}(min(\mu_X(x), \mu_Y(y))) \tag{6}$$

Figure 1 shows the fuzzy number: $\tilde{X} = (x, l_X, r_X) = (6, 4, 10)$. Left spread: $l_X = 4$. Right spread: $r_X = 10$. Most likely 6. $\tilde{X}$ can take values from 2 to 10. The most possible value (to the degree of 1) is 6. The possibility that the number $\tilde{X}$ will take a value in the range $[8, 11]$ is ½, and the possibility that it will take values from the range $[7, 8.8]$ is 3/4.

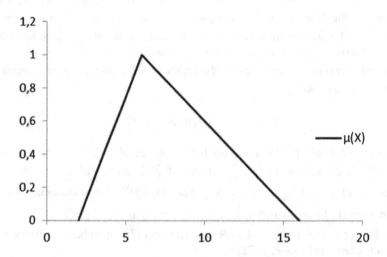

Fig. 1. Membership function of triangular fuzzy number $\tilde{X} = (x, l_X, r_X) = (6, 4, 10)$.

Suppose we wanted to compare the two fuzzy numbers. Let $\tilde{X} = (6, 4, 10)$ and $\tilde{Y} = (9, 5, 2)$, then the possibility that $\tilde{X}$ is greater or equal $\tilde{Y}$ is $Pos(\tilde{X} \geq \tilde{Y}) = 0.8$, see Fig. 2.

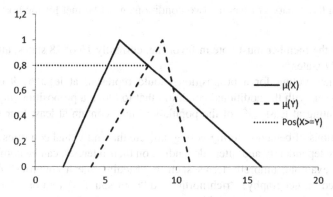

Fig. 2. Possibility that $\widetilde{X} \geq \widetilde{Y}$.

We will suggest the following fuzzy multi-criteria game (using many weights). Suppose we have several fuzzy weights (criteria, conditions) for each coalition $T \subseteq N$: $\widetilde{A}_k(T)$, $k = 1, \ldots, K$. The pair $\left(\widetilde{N, v}\right)$ will be called a fuzzy multicriteria simple cooperative game based on the set of players N and the real characteristic function v if for each coalition $T \subseteq N$ the characteristic function takes the value

$$v(T) = Pos\left(\bigcap_{k=1}^{K}\left(\widetilde{A}_k(T) \geq \widetilde{Q}_k\right)\right) \tag{7}$$

where:

$\widetilde{A}_k(\mathrm{T}) = (m_{\mathrm{T}}, \alpha_{\mathrm{T}}, \beta_{\mathrm{T}})$ – the fuzzy weight for k-th criterion of the coalition T, $k = 1, \ldots, K$

$\widetilde{Q}_k = \left(q_k, \alpha_{Q_k}, \beta_{Q_k}\right)$ - fuzzy quota for k-th criterion (in special cases the quota may be a crisp number), $k = 1, \ldots, K$.

4 Voting Procedure in the EU Council and Power Index of Groups of Member States

We conduct the analysis of changes in power relations between individual member states of the European Union based on decisions made in the EU Council of Ministers. Each country is represented in this Council by its representative who has the number of votes assigned to him in the vote.

Based on the Treaty of Lisbon [3] two conditions must be met for the normal voting to be valid:

(1) 55% of the member must vote in favor, i.e. currently 16 of 28 states, after Brexit, 15 of 27 states.
(2) The states voting for a proposition should represent at least 65% of the EU population, with the additional condition that to block a proposition any coalition representing at least 35% of the population must contain at least four states.

Observations of behavior during voting indicate that individual countries do not act completely independently and often, depending on their interests, can be combined into groups. For example, umbrella terms such as "Mediterranean countries" or Eastern Europe (based on geography), "rich north" and "poor south" (based on economics) or "old" and "new" Europe (based on history) are often used. Mercik and Ramsey (2017) made the division based on 8 selected economic indicators and created the following clusters (the leading country is listed first, followed by a number representing the number of countries and the percentage of the population of the EU that the cluster represents after Brexit)[4]:

Germany, Sweden, Austria, Denmark, Luxemburg (5; 23.83)
France, Netherlands, Belgium, Finland, Ireland (5; 23.61)
Italy, Portugal, Cyprus (3; 16.24)
Spain, Greece, Croatia (3; 13.86)
Poland, Romania, Czechia, Hungary, Bulgaria, Slovakia, Lithuania, Slovenia, Latvia, Estonia and Malta (11; 22.46)

Table 1 contains the results of the Shapley-Shubik power index calculation for each of these groups before and after Brexit.

Table 1. Shapley-Shubik power index for clusters around given member states pre and after Brexit (data taken from Mercik and Ramsey 2017)

	Germany (D)	France (F)	UK (UK)	Italy (I)	Spain (E)	Poland (PL)
Pre-Brexit	10/60	10/60	7/60	7/60	7/60	19/60
After-Brexit	14/60	14/60	–	9/60	9/60	14/60

5 Power Index of Groups of Member States – A Fuzzy Attempt

As already mentioned, the classic approach to calculating the Shapley-Shubik power index assumes that each of the possible coalitions is equipossible. In this sense, previously considered clusters of member states grouped around leading countries can

---

[3] We will notice that although the authors of the treaty rather did not expect Brexit to happen, the decision-making rules were formulated to be work regardless of the number of Member States.
[4] You can also show that after Brexit Ireland will join the group led by France.

create any coalitions. We do not deny this fact, however, we think that treating the entire cluster as a monolith with a certain number of votes (weights) and the population is over-simplistic. In individual votings, however, we are dealing with countries, not groups of countries, and it is easy to imagine situations when their interests are not necessarily compatible with the cluster. However, such situations can be modeled in a fuzzy approach, when there is no absolute necessity to follow the leader.

The process starts with specifying two fuzzy sets for each of the following blocks (clusters): $\widetilde{A_s}(T)$ - describing the number of Member States belonging to a given cluster, and $\widetilde{A_P}(T)$ - describing the population of this cluster. We assumed in a natural way that considering the size of a given cluster, in each case it can change by ± 1, and the possibility of such a change is equal to ½ (possibility of change by $\pm 2, \pm 3, \pm 4, \pm 5$ respectively is equal 0.3, 0.2, 0.1, 0.05). It is harder to make a natural assumption with regard to the population size of a given cluster. We assumed that the population size of a given country can be described by a triangular fuzzy number, where the most possible value is the population of a given cluster, left spread equals the population of the smallest country of the given cluster, while right spread equals population of the largest country outside the given cluster, excluding the leading countries (both figures are expressed as a percentage of the population of the entire European Union)[5]. Fuzzy weights for individual coalitions of clusters are presented in Table 2.

In the next step, based on formulas (3) and (6), we make calculations of a characteristic function for individual coalitions $v(T) = Pos\left(\widetilde{A_S}(T) \geq 15 \; AND \widetilde{A_P}(T) \geq 65\right)$. Table 2 presents relevant results for all coalitions of clusters.

Table 2. Fuzzy weights and functions characteristic of individual coalitions of clusters.

Coalition T	$\widetilde{A_s}(T)$ - states	$\widetilde{A_p}(T)$	$v(T)$
{D}	4|0.5 + 5|1 + 6|0.5	(23.83, 0.13, 4.48)	0,00
{F}	4|0.5 + 5|1 + 6|0.5	(23.61, 1.04, 4.48)	0,00
{I}	2|0.5 + 3|1 + 4|0.5	(16.24, 0.19, 4.48)	0,00
{E}	2|0.5 + 3|1 + 4|0.5	(13.86, 0.95, 4.48)	0,00
{PL}	10|0.5 + 11|1 + 12|0.5	(22.46, 0.09, 3.82)	0,00
{D, F}	8|0.3 + 9|0.5 + 10|1 + 11|0.5 + 12|0.3	(47.44, 1.17, 8.96)	0,00
{D, I}	6|0.3 + 7|0.5 + 8|1 + 9|0.5 + 10|0.3	(40.07, 0.32, 8.96)	0,00

(*continued*)

---

[5] This approach can be referred to as optimistic approach. The pessimistic approach in this case means the Left spread equal to the population of the largest country from a given cluster (except the leading country), and the right spread equal to the population of the smallest of the other countries. Probably a mixed approach should be used, but we have no premises to propose such an approach.

Table 2. (*continued*)

Coalition T	$\widetilde{A_s}(T)$ - states	$\widetilde{A_p}(T)$	$v(T)$
{D, E}	6|0.3 + 7|0.5 + 8|1 + 9|0.5 + 10|0.3	(37.69, 1.08, 8.96)	0,00
{D, PL}	14|0.3 + 15|0.5 + 16|1 + 17|0.5 +18|0.3	(46.29, 0.22, 8.30)	0,00
{F, I}	6|0.3 + 7|0.5 + 8|1 + 9|0.5 + 10|0.3	(39.85,1.23, 8.96)	0,00
{F,E}	6|0.3 + 7|0.5 + 8|1 + 9|0.5 + 10|0.3	(37.47, 1.99, 8.96)	0,00
{F, PL}	14|0.3 + 15|0.5 + 16|1 + 17|0.5 +18|0.3	(47.07, 1.13, 8.30)	0,00
{I, E}	4|0.3 + 5|0.5 + 6|1 + 7|0.5 +8|0.3	(30.10, 1.14, 8.96)	0,00
{I, PL}	12|0.3 + 13|0.5 + 14|1 + 15|0.5 + 16|0.3	(38.70, 0.28, 8.30)	0,00
{E, PL}	12|0.3 + 13|0.5 + 14|1 + 15|0.5 + 16|0.3	(36.32, 1.04, 8.30)	0,00
{D, F, I}	10|0.1 + 11|0.3 + 12|0.5 + 13|1 + 14|0.5 + 15|0.3	(63.68, 1.36, 13.44)	0,30
{D, F, E}	10|0.1 + 11|0.3 + 12|0.5 + 13|1 + 14|0.5 +15|0.3	(61.30, 2.12, 13.44)	0,30
{D, F, PL}	18|0.1 + 19|0.3 + 20|0.5 + 21|1 + 22|0.5 + 23|0.3	(69.90, 1.26, 12.28)	1,00
{D, I, E}	8|0.2 + 9|0.3 + 10|10.5 + 11|1 + 12|0.5 + 13|0.3 + 14|0.2	(53.93, 1.27, 13.44)	0,00
{D, I, PL}	16|0.2 + 17|0.3 + 18|10.5 + 19|1 + 20|0.5 + 21|0.3 + 22|0.2	(62.53, 0.41, 12.78)	0,81
{D, E, PL}	16|0.2 + 17|0.3 + 18|10.5 + 19|1 + 20|0.5 + 21|0.3 + 22|0.2	(60.15, 1.17, 12.78)	0,62
{F, I, E}	8|0.2 + 9|0.3 + 10|10.5 + 11|1 + 12|0.5 + 13|0.3 + 14|0.2	(53.71, 2.18, 13.44)	0,00
{F, I, PL}	16|0.2 + 17|0.3 + 18|10.5 + 19|1 + 20|0.5 + 21|0.3 + 22|0.2	(62.31, 1.32, 12.28)	0,78
{F, E, PL}	16|0.2 + 17|0.3 + 18|10.5 + 19|1 + 20|0.5 + 21|0.3 + 22|0.2	(59.93, 2.08, 12.78)	0,60
{I, E, PL}	14|0.2 + 15|0.3 + 16|10.5 + 17|1 + 18|0.5 + 19|0.3 + 20|0.2	(52.56, 1.23, 12.28)	0,00
{D, F, I, E}	12|0.1 + 13|0.2 + 14|0.3 + 15|10.5 + 16|1 + 17|0.5 + 18| 0.3 + 19|0.2 +20|0.1	(77.54, 2.31, 17.92)	1,00
{D, F, I, PL}	20|0.1 + 21|0.2 + 22|0.3 + 23|10.5 + 24|1 + 25|0.5 + 26| 0.3 + 27|0.2	(86.14, 1.45, 13.86)	1,00
{D, F, E, PL}	20|0.1 + 21|0.2 + 22|0.3 + 23|10.5 + 24|1 + 25|0.5 + 26| 0.3 + 27|0.2	(83.76, 2.21, 16.24)	1,00
{D, I, E, PL}	18|0.1 + 19|0.2 + 20|0.3 + 21|10.5 + 22|1 + 23|0.5 + 24| 0.3 + 25|0.2 +26|0.1	(76.39, 1.35, 17.26)	1,00
{F, I, E, PL}	18|0.1 + 19|0.2 + 20|0.3 + 21|10.5 + 22|1 + 23|0.5 + 24| 0.3 + 25|0.2 +26|0.1	(76.17, 2.27, 17.26)	1,00
{D, F, I, E,PL}	22|0.05 + 23|0.1 + 24|0.2 +25|0.3 + 26|0.5 + 27|1	(100, 2.40, 0)	1,00

Note that since none of the coalitions can exceed the value 27 in relation to the size of the countries, in three cases (full coalition and {D, F, I, PL} and {D, F, E, PL}) it was necessary to accept such a value of right spread that the possibility of the weight of the coalition exceeding 27 would be 0. Similarly, the right spread was defined for the population for several four- and five-element coalitions.

Based on the characteristic function $v(T)$ we calculate the values of the Shapley-Shubik power index for individual clusters. Table 3 shows the calculated values.

Table 3. Shapley-Shubik power index after Brexit for clusters around given member: standard and fuzzy versions (standard values for pre-Brexit have been scaled to account for the absence of the UK and allow comparability with later results).

	Cluster				
	Germany (D)	France (F)	Italy (I)	Spain (E)	Poland (PL)
Standard (pre-Brexit)	0,189	0,189	0,132	0,132	0,358
Standard (after Brexit)	0,233	0,233	0,150	0,150	0,233
Fuzzy	0,232	0,228	0,139	0,106	0,297

6   Discussing the Results and Further Work

Table 3 shows the changes observed in Shapley values in the standard and fuzzy approach. Discussion of the obtained results should also consider the standard pre-Brexit values. We can conclude that:

- Comparison of pre- and post-Brexit Shapley values points to the strengthening of all clusters except the cluster built around Poland.
- Fuzziness significantly affects two clusters: the cluster around Spain (a radically smaller value) and the cluster around Poland (a radically higher value).
- The strategic alliance from Poland declared by Poland had an impact on Shapley values for both clusters. Brexit apparently lowered Shapley value for Poland. However, taking into account the fact that individual countries do not unite around the leaders (this is the essence of fuzziness) restores the high value for the cluster around Poland.

Note that the fuzziness in our model can also be interpreted as the cohesiveness of countries within a given cluster. We called this approach to fuzziness an optimistic approach, which means that the countries with divergent views (not following the leader) are, de facto, small countries. Their change of position in a given vote, it seems, should be insignificant. However, it is not true. For at least two clusters (Spain and Poland), this impact is significant. This result is surprising especially when we take into account the fact that the cluster around Poland is significantly more numerous. It seems that when a cluster seems to be weak (in terms of size) is not really such. This is a confirmation of the observation made by Alonso-Meijide et al. (2009), ranking small countries higher than their population would suggest.

This time we do not aim to discuss the decomposition of Shapley value within a given cluster. At the basis of this possible decomposition lies the so-called the second axiom formulated by Myerson (1977) concerning the formation of a coalition (cluster). According to this axiom, the relations between the players in the coalition are symmetrical, i.e. the benefits should be the same for both players forming the given precoalition (cluster). In the case of forming clusters in the game regarding the European Union Council, the fuzziness is not symmetrical (the left and right-side spreads are different) and therefore it is difficult to rationally distribute the Shapley value in the game within the cluster. Mercik and Ramsey (2017) propose a symmetrical, and therefore non-fuzzy, approach. Their final result (despite over a year old) remains unchanged: „Although the UK will leave the EU. it is quite possible that there will eventually be an agreement that the UK will participate in the European free market (in a similar manner to Norway and Switzerland). Due to the UK's size and economic power (particularly in the financial sector). It is likely that the UK will play an important role in developing trade policy, while relinquishing power in the fields of social and federal policy".

Literature

Alonso-Meijide, J., Bowles, C., Holler, M., Napel, S.: Monotonicity of power in games with a priori unions. Theory Decis. **66**: 17–37 (2009)

Bertini, C.: Shapley value. In: Dowding, K. (ed.) Encyclopedia of Power, Los Angeles, pp. 600–603. SAGE Publications (2011)

Borkotokey, S.: Cooperative games with fuzzy coalition and fuzzy characteristic function. Fuzzy Sets Syst. **159**, 138–151 (2008)

Borkotokey, S., Mesiar, R.: The Shapley Value of cooperative games under fuzzy setting: a survey. Int. J. Gen Syst **43**(1), 75–95 (2014)

van den Brink, R.: An axiomatization of the Shapley value using a fairness property. Int. J. Game Theory **30**, 309–319 (2001)

Dubois, D., Prade, H.: Operations on fuzzy numbers. Int. J. Syst. Sci. **9**(6), 613–626 (1978)

Dubois, D., Prade, H.: Possibility Theory: An Approach to Computerized Processing of Uncertainty. Plenum Press, New York (1988)

Gładysz, B., Mercik, J.: The Shapley Value in fuzzy simple cooperative games. In: Nguyen, N. T., Hoang, D.H., Hong, T.-P., Pham, H., Trawiński, B. (eds.) ACIIDS 2018. LNCS (LNAI), vol. 10751, pp. 410–418. Springer, Cham (2018). https://doi.org/10.1007/978-3-319-75417-8_39

Li, S., Zhang, Q.: A simplified expression of the Shapley function for fuzzy game. Eur. J. Oper. Res. **196**, 234–245 (2009)

Meng, F., Zhang, Q.: The Shapley value on a kind of cooperative games. J. Comput. Inf. Syst. **7**(6), 1846–1854 (2011)

Mercik, J., Ramsey, David M.: The effect of Brexit on the balance of power in the European Union Council: an approach based on pre-coalitions. In: Mercik, J. (ed.) Transactions on Computational Collective Intelligence XXVII. LNCS, vol. 10480, pp. 87–107. Springer, Cham (2017). https://doi.org/10.1007/978-3-319-70647-4_7

Myerson, R.: Graphs and cooperation in games. Math. Oper. Res. **2**(3), 225–229 (1977)

Owen, G.: Values of games with a priori unions. In: Henn, R., Moeschlin, O. (eds.) Mathematical Economics and Game Theory, vol. 141, pp. 76–88. Springer, Heidelberg (1977). https://doi.org/10.1007/978-3-642-45494-3_7

Pang, J., Chen, X., Li, S.: The Shapley values on fuzzy coalition games with concave integral form. J. Appl. Math. (2014). http://dx.doi.org/10.1155/2014/231508

Shapley, L.S.: A value for n-person games. In: Kuhn, H., Tucker, A.W. (eds.) Contributions to the Theory of Games II. Annals of Mathematics Studies, vol. 28, pp. 307–317. Princeton University Press (1953)

Shapley, L.S., Shubik, M.: A method for evaluating the distribution of power in a committee system. Am. Polit. Sci. Rev. **48**, 787–792 (1954)

Tsumari, M., Tetsuzo, T., Masahiro, I.: A Shapley function on a class of cooperative fuzzy games. Eur. J. Oper. Res. **129**, 596–618 (2001)

Young, H.: Monotonic solutions of cooperative games. Int. J. Game Theory **14**, 65–72 (1985)

Yu, X., Zhang, Q.: An extension of fuzzy cooperative games. Fuzzy Sets Syst. **161**, 1614–1634 (2010)

Zadeh, L.A.: Fuzzy sets. Inf. Control **8**, 338–353 (1965)

Robustness of the Government and the Parliament, and Legislative Procedures in Europe

Chiara De Micheli[1] and Vito Fragnelli[2]([⊠])

[1] Department of Political Sciences, Communication Sciences and Engineering of Information, University of Sassari, Sassari, Italy
cdemicheli@uniss.it
[2] Department of Science and Innovative Technologies, University of Eastern Piedmont, Alessandria, Italy
vito.fragnelli@uniupo.it

Abstract. In a previous paper (see [3]), we analyzed the procedures of the Italian Constitution, focussing on their strength correlating it with the strength of the government and of the Parliament, measured through two parameters, the governability and the fragmentation. Here, we extend the analysis to other European democracies: United Kingdom, France, and Spain.

Keywords: Legislative procedures · Governability · Fragmentation Robustness

1 Introduction

In [3], we hypothesized and empirically verified that law makers are able to guide decision-making processes towards the maximization of their utility through the strategic use of procedural tools, referring to the Italian situation, thanks to high quantity and quality of data in our availability. The Italian case shows that often the government balances the weak control of its parliamentary majority using particular decision-making procedures. The executive extends both quantitatively (e.g. using frequently the decentralized commission procedure) and/or qualitatively (e.g. stretching the concepts of necessity and urgency) the application of some regulatory procedures when the uncertainty about the behavior of the majority is over a reasonable threshold.

By these means the executive strengthens its ability to impose on the Parliament.

In this paper, we aim to extend the research to other European countries, carrying out a comparative analysis. For this purpose, three parliamentary democracies, namely United Kingdom (UK), France and Spain, which differ in terms

N. T. Nguyen et al. (Eds.): TCCI XXXI, LNCS 11290, pp. 92–106, 2018.
https://doi.org/10.1007/978-3-662-58464-4_9

of party system and age of democratic consolidation will be compared with the Italian one[1].

The English democracy is the most consolidated and its process of democratization has never been interrupted. It is characterized by a highly organized and institutionalized party system. Conversely, France experienced a more discontinuous democratization process and its party system is composed by parties weakly organized and poorly rooted at the territorial level. Among those investigated, Spanish democracy is the youngest and the less consolidated. In the Spanish party system there are many regional political formations, strongly organized at the territorial level, which, in many cases, negotiate and give their parliamentary support to governments.

We want to make clear that, as in [3], we followed an experimental approach, looking at the evidence of when and how the different procedures were adopted in the legislative process in the various legislatures and comparing it with the robustness of the government and of the Parliament measured via the governability and the fragmentation.

The paper is organized as follows: in Sect. 2, we recall the governability and the fragmentation indices used for measuring the strength of the Government and of the Parliament; Sect. 3 is devoted to a description of the standard and non-standard procedures in the four countries; in Sect. 4, we present the tables with the data of the independent and dependent variables for each country; in Sect. 5, we perform a comparative analysis of the data and verify if the hypotheses formulated in [3] are still valid; Sect. 6 concludes.

2 The Independent Variables: Indices of Governability and Fragmentation

In this section we shortly recall the governability and the fragmentation indices (independent variables) that allow evaluating the robustness of the Government and of the Parliament, addressing to [3] for further details.

The *governability* of the Parliament is the capability of the parties to form a strong majority. It is inversely related to the number of parties and directly related to the number of seats of the majority.

The first governability index (see [5]) considers the number m of critical parties, i.e. those parties that may destroy the majority withdrawing, and the number of seats of the majority, f.

The formula for the governability index g_1 is:

$$g_1 = \frac{1}{m+1} + \frac{1}{m(m+1)} \frac{f - T/2}{T/2}$$

where T is the total number of seats.

---

[1] Spain reached democratization in the period called *third wave* by Huntington [6] that began in 1974; UK and France democratized in the *first wave*, from 1828 to 1926, and Italy in the *second wave*, from 1943 to 1962.

It is possible to remark that for each value of m we have $\frac{1}{m+1} \leq g_1 \leq \frac{1}{m}$. For instance, if the government is supported by just one critical party, g_1 is in between 0.5 and 1; if it is supported by two parties, then g_1 is in between 0.333 and 0.5, and so on. The value of g_1 in the range determined by m depends on the number of seats of the majority coalition and on the total number of seats.

In order to increase the importance of the number of seats of the majority coalition, we consider a second governability index g_2 in which we take into account the percentage of seats of the majority divided by the number of parties in the majority (see [7]):

$$g_2 = \frac{f/T}{p}$$

where f and T are the same as above and p is the number of parties in the majority.

The *fragmentation* of the Parliament is a measure that accounts not only the number of parties, or groups, or factions that compose the Parliament, but also their number of seats.

In order to analyze the role of factions, i.e. the fragmentation, we refer to the classical index of Rae-Taylor [12], defined as:

$$I_{RT} = 1 - \sum_{i \in N} s_i^2$$

where $N = \{1, ..., n\}$ is the set of parties and s_i is the percentage of seats of party $i \in N$.

It assumes values in the range $[0, 1]$, and the value mainly depends on large parties, i.e. very small parties do not affect too much the fragmentation.

The following example shows the behavior of the indices g_1, g_2 and I_{RT}.

Example 1 (see Example 1 in [3]). *Consider a Parliament of 100 seats; suppose that the majority includes parties P1, P2, P3 and P4; consider three situations s1, s2 and s3 in which the seats of these parties are distributed according to the following table (critical parties are in **bold**), where we compute also the three indices we are interested in:*

	P1	P2	P3	P4	f	m	g_1	g_2	I_{RT}
s1	**20**	**16**	**14**	**13**	63	4	0.2130	0.1575	0.7428
s2	**17**	**17**	**17**	11	62	3	0.2700	0.1550	0.7430
s3	**49**	2	2	2	55	1	0.5500	0.1375	0.2023

g_1 and g_2 *behave in opposite ways, the former increases, the latter decreases, while* I_{RT} *first increases and then decreases.*

3   The Dependent Variables: Decision-Making Procedures

In this section, we analyse the decision-making procedures (dependent variable) used in the four countries. The aim is to verify if, how and to what extent governments make use of the opportunities offered by the different institutional instruments to produce laws. Each country is examined following descriptive criteria.

3.1   United Kingdom

There are two chambers, the House of Lords (*Upper chamber*) and the House of Commons (*Lower chamber*). A law has to be approved by the two chambers, but the Lower chamber is more important; in fact, it is the sole that has to approve the financial laws; moreover, the House of Commons can override the objections of the House of Lords and pass a law without the consent of the latter. The Standing Committees have the role of examining and amending the proposals.

The fundamental normative instruments used by the government are the so-called *statutory instruments*. The Statutory Instruments Act of 1946 enables UK members of government to enact a sort of delegated legislation grouping different procedures, e.g. partially similar to the delegating laws in the Italian Constitution. Those procedures may provide substantive regulatory acts (as the Regulation or the Deregulation Orders) or, in other cases, may provide formal regulatory acts (as the Commencement Orders) through which the government, delegated by the Parliament, establishes when a certain legislative act comes in force. All those legislative instruments have to be approved by the individual ministers, in their turn delegated by a parliamentary enabling act. In essence, the procedure runs as follows: a statutory instrument is approved and presented by a minister and will be effective after 40 days, within which a parliamentary control could be activated and the instrument could be annulled by a negative resolution of the Parliament. It can be observed that, if compared with those contemplated by the Italian Constitution, the UK governmental legislative instruments lack the urgency requirement. Nevertheless, when an urgency has to be faced, the statutory instrument runs immediately in force before its presentation, after communication to the Lord Speaker[2] and to the Speaker of the Commons, and the parliamentary control could be considered as a following act. In this case, the statutory instrument could lose its force if a negative resolution is enacted by the Parliament within 40 days; if no resolution is adopted, it maintains its force. In few cases an opposite situation is expected: when the enabling act expressly establishes that the statutory instrument is in force if the Parliament votes an affirmative resolution within a certain time limit (see [9]).

---

[2] Until 2005, the president of the House of Lords was Lord Chancellor, and was the person to whom statutory instruments were communicated.

3.2   France

There are two chambers, the *Senat* (*Upper chamber*) that is less important than the *Assemblée Nationale* (*Lower chamber*) as the government may ask for its final vote in case of disagreement, after two rounds (art. 45). In some cases, it is possible a faster procedure with just one round in each chamber. The laws are discussed only in the permanent committees (presently 8), a part from the financial law, the constitutional revisions and the financial supports to the social security.

We have investigated the use of the so-called *ordonnances* (ordinances) that can be adopted according to art. 38 of the French Constitution[3]. Another kind of procedure is the confidence vote, used by the French government to force the favorable vote of the parliamentary majority on a given issue.

As art. 49 paragraph 3 of the French Constitution states: *The Prime Minister may, after deliberation by the Council of Ministers, make the passing of a Finance Bill or Social Security Financing Bill an issue of a confidence vote of the National Assembly. In that event, the Bill shall be considered passed unless a resolution of no-confidence, scheduled within the subsequent twenty-four hours, is carried out as provided for in the foregoing paragraph. In addition, the Prime Minister may use the above procedure for one other Government or Private Members' Bill per session*[4] (see [10]).

3.3   Spain

The Parliament (*Cortes Generales*) includes the *Senado* (*Upper chamber*) and the *Congreso de los Diputados* (*Lower chamber*) (art. 66). The Lower chamber is more important as the role of the Upper chamber is limited to the possibility of a veto, that requires an absolute majority, or amending a law. The Lower chamber may delegate the committees (art. 75), but it is possible to take back upon itself the discussion and the approval of any law, in any moment (art. 78).

---

[3] With a view to carrying out its programme, the Government may seek the authorization of Parliament, for a limited period of time, to issue ordinances regulating matters normally falling within the field of law-making. (The) ordinances are made in the Council of Ministers after consultation with the Council of State. They come into force upon publication, but cease to be effective if the Bill ratifying them is not laid in front of Parliament by the date fixed by the enabling Act. At the expiration of the period mentioned in paragraph 1 of this article, (the) ordinances may be modified only by law, as regards matters falling within the field of law.

[4] This is a *passive* confidence, differently from the Italian one that may be viewed as an *active* confidence. More precisely, the difference refers to countries where the government needs to win an investiture vote, and countries in which the government just needs to be tolerated by Parliament. For further details see [14].

The 1978 Spanish Constitution disciplines both *delegación legislativa* (legislative delegation, art. 82)[5] and *decreto-ley* (decree law, art. 86)[6] procedures. The remarkable use of the decree laws became, in certain periods, a feature of Spanish law-making, partially due to a particular constitutional discipline (art. 86) providing the prolongation of effectiveness of the decree law, even if a parliamentary vote of conversion (the so-called *convalidacion*) does not occur (see [8]). The vote of conversion has to take place within 30 days[7]. Usually, less than 30% of Spanish decree laws have been converted[8]. For this reason all the enacted decree laws, even if not converted are taken into account here.

3.4 Italy

A short recall of the Italian situation can be useful, again addressing to [3] for further details. The Parliament is formed by two chambers the *Senato della Repubblica* (Upper chamber) and the *Camera dei Deputati* (Lower chamber) with equal role in this matter. According to the Italian standard procedure, any member of the Parliament can propose a bill, not only the Government; a committee (*referral commission*), often integrated with an executive's representative as observer, discusses and amends it. Eventually, the two chambers examine the bill and vote it article by article and in full; the two chambers have to approve the same text. It can be used to approve all types of bill but, for some of them, a special quorum is required.

If compared with the other three democracies, the Italian Constitution provides a greater number of procedures to legislate. The most important, both quantitatively and qualitatively, are the following.

In the period of democratic establishment and consolidation, there was an exasperated use of the *procedura decentrata* (decentralized procedure, art. 72, c.2)

---

[5] The Cortes Generales may delegate to the Government the power to issue rules with the force of an act of the Parliament on specific matters not included in the foregoing section. Legislative delegation must be granted by means of act of basic principles when its purpose is to draw up texts in sections, or by an ordinary act when it is a matter of consolidating several legal statutes into one.

[6] In case of extraordinary and urgent need, the Government may issue temporary legislative provisions which shall take the form of decree laws and which may not affect the legal system of the basic State institutions, the rights, duties and freedoms of the citizens contained in Part 1, the system of Self-governing Communities, or the general electoral law. Decree laws must be inmediately submitted for debating and voting by the entire Congress, which must be summoned for this purpose if not already in session, within thirty days of their promulgation. The Congress shall adopt a specific decision on their ratification or repeal in the said period, for which purpose the Standing Orders shall provide a special summary procedure. During the period referred to in the foregoing subsection, the Cortes may process them as Government bills by means of the urgency procedure.

[7] For instance, the decree law n. 5 of 24 May 2002 was cancelled, as no necessity and urgency reasons hold.

[8] For instance, the decree laws were the 27% of the activity in the V legislature and the 28% in the V legislature.

according to which the legislative process (final approval included) takes place only in a Committee, if there is consensus among political actors, otherwise the bill will return to the ordinary procedure.

The Italian constitution, such as the Spanish one, although with differences (see above) also allows the *decreto legge* (decree law, art. 77, c.2) that enables the Executive issuing a decree in "extraordinary cases of necessity and urgency". The Executive's decree becomes law immediately and remains in effect for 60 days without any parliamentary approval. If after this period the Parliament has not converted the decree into a perfect law, then previous status quo is re-established. The unique procedure to convert a decree law is the ordinary legislative procedure, so due to the deadline of 60 days, the members of the Parliament often bargain with the Executive in order to add new normative contents. After the sentence of the Constitutional Court in 1996, a new decree may reiterate a decayed decree only if the following conditions occur: the government founds the reiteration of the decree on new arguments about his extraordinary necessity and urgency; the government characterizes the contents of the reiterated decree with different regulatory arrangements.

In the Second Republic (from 1994 on) a procedure provided by the Constitution obtains centrality: the *legge delega* (delegating law, art. 76 and art. 77, 1). The delegating law is approved by the standard procedure. This type of bill has at least a section delegating to the Executive the power to promulgate the legislative decrees according to some general framework voted in the delegating law, and within a limited period of time. The legislative decree approved by the Council of Ministers is sent to the President of the Republic, at least 20 days before the deadline required by the delegating law, so that the President can check it and, if necessary, send it back to the Chambers.

Finally, there is a strategic use on the confidence vote in order to force the fast approval of a law, possibly without modification or with a unique amendment, the so-called *maxi-emendamento*, reducing the vulnerability of a law approved article by article. The confidence vote is an extreme and effective attempt to protect the content of a Government bill. It is worth mentioning that the government remains in office until enjoys the confidence of the parliamentary majority, whose existence can be verified at any time.

4   The Data

In this section, we present the data we used for measuring the trend of the independent and dependent variables referring to both qualitative and quantitative data. First, in each country the robustness of the Parliament and the government (the independent variables) is assessed through the two governability indices and the fragmentation index. Then, the law making procedures used in the legislative processes of the four countries (the dependent variables) are analysed.

The *terminus ad quem* of the analysis is 2016 for all the four countries, while the *terminus a quo* is different for each democracy: 1974 for UK, 1958 for France, 1993 for Spain and 1948 for Italy.

4.1   Governability and Fragmentation

In this subsection a diachronic analysis of each country and a comparative analysis of the four countries will be conducted regarding the governability degree and the fragmentation degree of their government-Parliament sub-system.

The data for the Italian Parliament represent the average value among the two chambers, while for the other countries they refer to the Lower Chamber that is the most important among the two. For each country, we report the number of the legislature (Leg), the starting year $(year)$, the two governability indices $(g_1$ and $g_2)$ and the fragmentation index (I_{RT}).

UK

Leg	XLVII	XLVII	IL	L	LI	LII	LIII	LIV	LV	LVI
year	1974	1979	1983	1987	1992	1997	2001	2005	2010	2015
g_1	0.503	0.534	0.611	0.577	0.516	0.636	0.627	0.552	0.353	0.508
g_2	0.503	0.534	0.611	0.577	0.516	0.636	0.627	0.552	0.280	0.508
I_{RT}	0.555	0.534	0.521	0.541	0.558	0.526	0.535	0.591	0.611	0.599

The English Parliament appears to be the less fragmented, while the governability degree of the UK governments is, in general, the highest. However, we can observe that the last three elections have produced a more fragmented Parliament and less governable cabinets (as an example, from 2010 to 2015, a coalition cabinet is formed, an extraordinary event for the English political system).

France

Leg	I	II	III	IV	V	VI	VII	VIII	IX	X	XI	XII	XIII	XIV
year	1958	1962	1967	1968	1973	1978	1981	1986	1988	1993	1997	2002	2007	2012
g_1	0.426	0.198	0.332	0.772	0.334	0.355	0.625	0.332	0.471	0.440	0.259	0.629	0.580	0.548
g_2	0.185	0.279	0.249	0.298	0.266	0.283	0.458	0.248	0.477	0.409	0.184	0.628	0.290	0.431
I_{RT}	0.718	0.715	0.660	0.431	0.682	0.770	0.650	0.753	0.700	0.672	0.738	0.566	0.596	0.644

The fragmentation degree of the French Fifth Republic (from 1958 to 2016) maintains high values – more than 0.7 – until 1993 (five legislatures), while it decreases in the following legislatures to values lower than 0.6. The governability of French cabinets varies between 0.4 and 0.6 and is lower than both UK and Spanish governability. In particular, the governability is affected by the so-called *cohabitation*, i.e. when presidential and parliamentary majorities are different; this happened during the VIII and XI legislatures characterized and for part of the X legislature[9].

---

[9] During 1986–1988, the *cohabitation* involved the socialist President François Mitterrand and the neo-Gaullist Premier Jacques Chirac; during 1993–1995, the cohabitation involved the socialist President François Mitterrand and the neo-Gaullist Premier Édouard Balladur; during 1997–2002, the cohabitation involved the neo-Gaullist President Jacques Chirac and the socialist Premier Lionel Jospin.

Spain

Leg	COS	I	II	III	IV	V	VI	VII	VIII	IX	X	XI	XII
year	1977	1979	1982	1986	1989	1993	1996	2000	2004	2008	2011	2015	2016
g_1	0.474	0.336	0.578	0.526	0.500	0.454	0.446	0.523	0.469	0.483	0.531	0.351	0.391
g_2	0.474	0.255	0.578	0.526	0.500	0.454	0.446	0.523	0.469	0.483	0.531	0.351	0.391
I_{RT}	0.654	0.641	0.573	0.626	0.648	0.625	0.632	0.595	0.598	0.571	0.613	0.757	0.737

For Spain (including the Constituent assembly, COS), we can observe a certain stability of fragmentation (in general 0.6) and governability (in general 0.5) until the X legislature, and a sharp change in the last two legislatures (2015[10] and 2016 general elections), when fragmentation increases over 0.7 and governability decreases below 0.4.

Italy

Leg	I	II	III	IV	V	VI	VII	VIII	IX	X	XI	XII	XIII	XIV	XV	XVI	XVII
year	1948	1953	1958	1963	1968	1972	1976	1979	1983	1987	1992	1994	1996	2001	2006	2008	2013
g_1	0.613	0.469	0.478	0.383	0.378	0.390	0.534	0.421	0.365	0.364	0.559	0.522	0.416	0.242	0.260	0.645	0.440
g_2	0.158	0.238	0.130	0.181	0.187	0.095	0.131	0.097	0.179	0.088	0.087	0.092	0.050	0.087	0.067	0.131	0.074
I_{RT}	0.704	0.708	0.692	0.731	0.736	0.735	0.701	0.716	0.762	0.762	0.824	0.857	0.846	0.838	0.826	0.711	0.770

The Parliaments of previous countries are less fragmented than the Italian Parliament, whose fragmentation is steadily over 0.7, and from the XI to the XV legislatures stands over 0.8. Moreover, in UK, France and Spain the level of governability is higher than in Italy, where governability rarely exceeds the value of 0.6 for each legislature.

4.2   Legislative Procedures

In this subsection, the ordinary legislation and parliamentary originated legislation are quantitatively examined and compared with the other forms of normative production.

Due to the high difference of durations of the legislatures, the data are reported per month, so we added also the number of months of each legislature ($\#m$).

---

[10] In 2016 for the first time, the Spanish Parliament was dissolved due to the impossibility of forming a majority; according to the Constitution, the King dissolves the Parliament if the designated Premier does not obtain the confidence vote in two months.

UK

Leg	XLVII	XLVII	IL	L	LI	LII	LIII	LIV	LV	LVI
#m	55	49	48	58	61	49	47	60	60	8
SP	5.38	5.35	5.29	4.47	4.69	3.31	4.74	2.72	2.23	3.00
SI	102.56	88.61	102.48	135.28	159.57	160.08	160.85	159.03	266.42	173.50

Ordinary laws (SP) and statutory instruments (SI) per month.
Source: http://www.legislation.gov.uk/uksi

In UK, the ordinary legislative production, during the period 1972–2016, consists of 4.2 laws per month, falling down from 1997 onward; as in the other countries the most recent legislatures are characterized by a strong reduction. Parliamentary originated legislation amounts in general to 16%. From 1993 to 2008 the statutory instruments enacted by the government remain constantly on 2000 per year, reaching a peak of 2285 in 2001. As far as the use of statutory instruments is concerned, from 2010 onward, during the Cameron cabinet, increased until 2014, when they reached their maximum number of 266.42 per month then falling down at 173.50 per month in the LVI legislature.

France

Leg	I	II	III	IV	V	VI	VII	VIII	IX	X	XI	XII	XIII	XIV
#m	47	48	52	15	57	60	39	57	27	57	51	60	60	60
SP											9.00	7.73	8.48	5.90
Ord	0.89	0.10	0.67	0.93	0.32	0.53	1.03	0.14	1.07	0.16	1.61	2.58	2.08	2.92
Con	0.09	0.04	0.08	0	0.14	0.10	0.13	0.56	0.56	0	0.02	0.03	0	0.07

Ordinary laws (SP) Ordinances (Ord), and confidence vote (Con) per month.
Source: http://www.assemblee-nationale.fr

The use of the ordinances shows a cyclical trend; it reaches the top in the last three legislatures when, on average, more than 2 ordinances per month were issued. This procedure has been used as a preferential lane to legislate on specific policy areas, such as: the transposition of the EU laws; the regulation of local authorities; and, more recently, regulatory simplification. The ordinary laws enacted by the Assemblée Nationale, for which data are available only for the last four legislatures, fall down steadily, from an average of 9.00 to 5.90. Parliamentary originated legislation has varied between a minimum of 9% and a maximum of 21%, placing yearly on an average of 17%.

Spain

Leg	COS	I	II	III	IV	V	VI	VII	VIII	IX	X	XI	XII
#m	20	32	44	28	44	33	48	48	48	44	49	6	17
SP	5.00	5.56	4.59	3.18	2.75	4.39	4.63	4.19	3.48	2.07	3.16	10.67	0.53
LD	2.80	1.78	1.32	0.61	0.55	1.42	1.63	0.90	1.06	0.86	1.55	0.17	1.47
DL	0	0.09	0.02	0.64	0.11	0.12	0.04	0.15	0.23	0.07	0.08	0	0.06

Ordinary laws (SP), legislative decree (LD) and decree law (DL) per month.
Source: http://www.congreso.es/portal/page/portal/Congreso/Congreso

On average ordinary legislation counts 3.4 laws per month, although there is a strong fluctuation over the years: 5.56 in the I legislature, 2.07 in the IX legislature and 0.53 in the XII legislature. About 10% of this ordinary legislation originates in the Parliament. Decree laws vary during the legislatures, never going over 1 per month.

Italy

Leg	I	II	III	IV	V	VI	VII	VIII	IX	X	XI	XII	XIII	XIV	XV	XVI	XVII
#m	62	57	59	59	49	48	35	45	49	56	22	24	60	57	22	55	42
SP	8.77	6.86	6.31	5.81	3.22	4.27	5.70	5.79	4.24	5.52	1.55	0.83	3.22	2.19	1.41	1.44	1.17
PD	25.50	22.07	20.30	19.56	11.15	12.94	5.70	9.86	6.37	6.75	4.41	1.33	4.15	2.28	0.32	1.11	0.33
DL	0.47	1.05	0.51	1.59	1.41	2.58	4.77	6.11	6.27	8.39	23.73	24.04	7.63	3.96	2.32	2.24	1.93
LD	0	0	0	0	0	0	0	0.18	0.14	0.46	0.73	0.25	1.10	0.86	0	0.60	0.57
F	0.08	0.23	0.07	0.39	0.08	0.13	0.09	0.07	0.69	0.50	0.82	1.08	0.45	0.72	1.00	1.40	1.02

Ordinary laws (SP), decentralized procedures (PD), decree law (DL), delegating law (LD) and confidence vote (F) per month.
Source: http://www.camera.it and http://www.senato.it

The common feature of all the Italian legislatures is the decreasing of the legislative production; including all the instruments, the number of laws goes down from about 35 per month to about 5 per month. The decentralized procedures are frequently used until the VIII legislature, when the decree laws became more used, until the XIII legislature, when the sentence of the Constitutional Court in 1996 made difficult to reissue a decree law, increasing the use of the confidence vote, possibly adding a maxi-amendment.

5   Comparative Analysis and Hypotheses Check

In this section, our aim is to verify through a comparative analysis of the four countries if, how and to what extent governments use the opportunities given by different institutional instruments to orient and/or control decision-making, balancing their weak control of the parliamentary majority. For each country, the use of the different procedures of law-making already discussed in Subsect. 4.2 are compared with the levels of both governability and fragmentation discussed in Subsect. 4.1.

The first remark is the confirmation of the reduction of the legislative production for all the four democracies; in particular, the governments expropriated the Parliaments of their normative duties. More precisely, the French Fifth Republic registered a reduction with respect to the Fourth Republic (see [13]); in the UK, the decreasing of the laws approved with the standard procedure, started at the beginning of the XX century, continues in the period under investigation, from 5.38 laws per month to less than 3 laws per month (see [15]); the legislative production in Spain is relatively more stable than in the other countries, with the exception of the current legislature that started only 17 months ago (at December 2017), so that this datum is not comparable with the others; finally, in Italy the reduction is more evident only in the last legislatures[11], but the Italian democracy remains one of the most productive among the Western ones (see [1, 11]).

UK

There are few formal limits to the decisional capacity of the government when it is supported by a cohesive one-party majority. This was an unchanging feature of the English political system until the 2001–2005 legislature[12], when the number of members of the Parliament who rebelled to the party discipline was the highest since 1945. Furthermore, the average degree of fragmentation remained constant till 2015, while the average degree of governability remained constant till 2010, then decreased (from 2010 to 2015 UK was governed by an unusual coalition between the Tories and the Liberal-democrats) and the use of statutory instruments sharply increased (with the highest production per month, more than 266); on the other hand, the use of the standard procedure further decreases. It is possible to suppose that the decreasing is, at least for the most recent governments, due to the lack of a cohesive and disciplined majority. Although this clear relationship between variables occurs only in the LV and LVI legislatures (those elected in 2010 and 2015) the link appears strong and empirically sustainable.

France

The degree of fragmentation is high, although lower than in Italy, and partially explains the use of confidence vote by the governments: it increases when parliamentary majorities are more fragmented or tight. Moreover, the confidence vote recurs more frequently when the high parliamentary fragmentation coex-

---

[11] Other legislative systems, even if they are enough different, as Germany, Japan, Norway and USA, show a decreasing in the legislative production after World War II [13]. The German democracy is the most productive, i.e. the closest to the Italian situation [4].

[12] There was an exception with the Callaghan government (1976–1979) during the XLVII legislature, when for most part of the term the government lacked a majority and had to rely on minor parties to govern.

ists with low governability (as in the V, VI and VII legislatures) or with high fragmentation of parliamentary majorities, and/or with minority governments (as it were the cases of socialist governments in the VI and IX legislatures). The use of the confidence vote tends to decrease in recent years, though it should be emphasised that constitutional amendments of 2008 strictly narrowed its application to the financial laws and to the laws on social security financing. During the XIV legislature, this procedure has been resumed and this may suggest the emergence of a new practice in the use of the confidence vote for legislative purposes. The use of ordinances increases when ordinary laws decrease, likewise the contrary occurs as in the XIII and XIV legislatures, jointly with the reduction of the use of confidence vote. It is also plausible for the ordinances, the hypothesis that their use increases when it is difficult for the majority to enact laws through ordinary legislation, as during the I, VII IX, XI and XIV legislatures, when the Parliament had a high degree of fragmentation. Even in this country, where the government, in the dual figure of the Prime minister and the President of the Republic, has broad chances to impose its own political agenda, in times of high fragmentation and low governability, the government itself adopts more frequently different procedures to control the recalcitrant parliamentary majority.

Spain
The degrees of fragmentation and governability are generally stable and placed around intermediate values till the X legislature. As a consequence, decree laws place steadily around intermediate values. It is worthwhile to remark that during the Constituency assembly, due to the urgency of building a new state there was a very high number of laws. This correlation is validated by the events that took place in the XI and XII legislature: increasing in the fragmentation and decreasing in the governability occurred simultaneously, making difficult a highest use of both decree laws and delegating legislation; this matter may be connected to the larger difficulties in making any agreement on basic interventions, leading to hard application of whatever procedure. In the VIII legislature, the government made an extensive use of legislative decrees, and from this legislature onward ordinary legislation has been difficult to be enacted by all the governments.

Italy
As far as the Italian democracy is concerned, previous empirical analysis (see [2,3]) showed that robust correlations exist between the following phenomena: (a) the extensive use of the decentralised deliberating procedure and a medium fragmentation degree; this procedure tended to decline when fragmentation increased; (b) the wide use of decree laws, in particular since the VII legislature, growing when the fragmentation increases; (c) although remarkable, a more discontinuous use of delegating laws starting from X legislature and growing during the so-called Second Republic, when fragmentation increased; (d) the confidence vote is widely used since IX legislature onward, and is strongly correlated with the increasing of fragmentation.

6   Concluding Remarks

The first comment concerns the continuous decline of ordinary legislation enacted in all examined countries. Albeit with some differences, in all these democracies law making processes are characterized by an increasing of the governmental normative instruments use, which substitute ordinary legislation when fragmentation increases and/or governability decreases. This trend is particularly evident in the Italian situation, that may be viewed as a good ground for testing the existence of connections among our independent and dependent variables. The comparative analysis allows deepening if a relation could be hypothesized between institutional and political weaknesses. The use of particular tools and decision-making procedures causes significant effects on the decision-making capacity of governments.

The "shortcut" of the procedures is, in the short term, the most attractive and probably more profitable strategy to increase the decision-making capacity of governments, if the outcomes of the ordinary legislative process are endangered by a high degree of fragmentation or low governability. However, this is a short-term strategy that has only a marginal impact on the real crucial issue of parliamentary systems, which is the behavior and discipline of parties within the Parliament.

Finally, it is possible to suppose that this behavior of the governments for monopolizing the legislative process is becoming a specific and permanent characteristic of how the subsystem government-Parliament works in this new century.

Acknowledgement. The authors gratefully acknowledge two anonymous referees for their useful comments and detailed remarks, and the participants to the workshop "Quantitative methods of group decision making" held at the Wroclaw School of Banking in November 2017 for useful discussions.

References

1. Camera dei Deputati: Rapporti annuali sulla legislazione (2010, in Italian)
2. De Micheli, C.: Parlamento e governo in Italia. Partiti, procedure e capacità decisionale dal 1948 al 2013. Franco Angeli, Milano (2014, in Italian)
3. De Micheli, C., Fragnelli, V.: Robustness of legislative procedures of the Italian Parliament. Trans. Comput. Collective Intell. **23**, 1–16 (2016)
4. Di Porto, V.: Il Parlamento del bipolarismo. Un decennio di riforme dei regolamenti delle Camere. In: Il Filangieri, Rivista di Diritto Pubblico e Scienza della Legislazione, ARSAE (2007, in Italian). www.arsae.it/arsae/index
5. Fragnelli, V., Ortona, G.: Comparison of electoral systems: simulative and game theoretic approaches. In: Simeone, B., Pukelsheim, F. (eds.) Mathematics and Democracy. Studies in Choice and Welfare. Springer, Heidelberg (2006). https://doi.org/10.1007/3-540-35605-3_5
6. Huntington, S.P.: The Third Wave: Democratization in the Late Twentieth Century. University of Oklahoma Press, Norman (1993)

7. Migheli, M., Ortona, G.: Majority, proportionality, governability and factions. Working paper n. 138, POLIS, University of Eastern Piedmont, Alessandria (2009)
8. Naranjo De La Cruz, R.: La reteración de los Decretos Leyes in Italia y su analisis desde ordenamiento constitucional espagnol. Revista de Estudios Politicos, vol. 99, pp. 257–280 (1998, in Spanish)
9. Palici Di Suni Prat, E.: La funzione normativa tra governo e parlamento. Profili di Diritto Comparato, CEDAM, Padova (1988)
10. Palici Di Suni Prat, E., Comba, M., Cassella, M.: Le Costituzioni dei Paesi della Comunità Europea. Valaio Editore, Pavia (1993)
11. Pasquino, G., Pellizzo, R.: Qual è il parlamento più produttivo? Contribution of 3 June (2016, in Italian). http://www.casadellacultura.it
12. Rae, D.W., Taylor, M.: The Analysis of Political Cleavages. Yale University Press, New Haven (1970)
13. Rose, R.: Understanding Big Government: The Programme Approach. Sage, London (1984)
14. Russo, F., Verzichelli, L.: The adoption of positive and negative parliamentarism: systemic or idiosyncratic differences? Presented at the ECPR Joint Sessions of Workshops, Salamanca, 10–15 April 2014 (2014)
15. Van Mechelen, D., Rose, R.: Patterns of Parliamentary Legislation. Gower Publishing Limited, Aldershot (1986)

Should the Financial Decisions Be Made As Group Decisions?

Anna Motylska-Kuzma$^{(\boxtimes)}$

WSB University in Wroclaw, Wroclaw, Poland
anna.motylska-kuzma@wsb.wroclaw.pl

Abstract. *Objective:* The aim of the article is to verify the effectiveness of financial decision making as group decisions.

Methodology: The research consisted of a simple simulation game conducted with a group of students. There were 540 single games analysed, taking into account the decisions and their effectiveness with regard to the size of the decision group. The ANOVA, the dependent paired samples test as well as the independent samples test were used to verify hypothesis.

Results: The studies show that simple decisions about production and price are almost the same, regardless of size of the group but the capital structure differ significantly between the groups and between the individuals and the group. The analysis results also indicate differences in effectiveness of the decision making.

Originality/Value: The findings of research can be used to enhance the performance of the decision making process in companies, especially within the scope of finance.

Keywords: Group decisions · Finance · Effectiveness

1 Introduction

The effectiveness of group decision process has been an increasingly important organizational concern [9]. The growing demand for efficiency and flexibility made the organisations engage the teams to solve many problems which have normally been considered by the individuals [4, 13, 16]. This is based on the assumption that decisions made by groups composed from the members with diversified experience and characters will be more effective and better quality than those made by the individuals with their limits and routine [13, 20]. In practice, there are groups with better effectiveness in decision making process and groups with lower effectiveness in comparison to the individuals. The results are strictly connected with the composition of the group [9, 24]. However, there are decisions which require better knowledge and the decisions which require a quick solution. Thus not every decision should be made by the group.

The financial decisions in the company are a very specific type of decisions. The effectiveness of such decisions is mostly assessed by the economic results and they need strictly defined responsibility. Therefore, it is really interesting to know, if such decisions should be made by the group or by individuals. Thus, the main object of this article is to verify the effectiveness of financial decision making as a group decision.

© Springer-Verlag GmbH Germany, part of Springer Nature 2018
N. T. Nguyen et al. (Eds.): TCCI XXXI, LNCS 11290, pp. 107–117, 2018.
https://doi.org/10.1007/978-3-662-58464-4_10

The author wants to find the answer to following questions: (1) Is the effectiveness of financial decision making the same, regardless of the size of the group; (2) Are the decisions made by the groups the same regardless of the size of the group; and finally (3) How the group behaves when the last decision was unsuccessful and if is it dependent on the size.

The paper is structured in a manner as follows. The first section explains the nature of financial decisions, their basic types, problems and ideas. The second section shows the characteristics of groups as a decision maker. The third and fourth sections describe the methodology of research and the findings. The final section presents the summary and the future directions of the validation of the research.

2 Financial Decisions in the Enterprises

Every decision made in a business has financial implications, and any decision that involves the use of money is a corporate financial decision. Nevertheless, the financial decisions are divided into three main categories [5]:

- Financing decisions – i.e. all decisions connected with optimization of capital structure, which translates into the effective engaging the debt and equity capital as well as the internal and external sources of capital in financing the activity of a company. The optimal capital structure is understood as such a leverage ratio which ensures the biggest profitability at the lowest risk. Every entrepreneur who operates in imperfect capital markets where there are transactions costs, taxes, bankruptcy and agency costs, and each of them makes determination of optimal capital structure, has to solve the question in order to maximize the value of the company he or she owns or manages [1]. Therefore, in the decision making process about financing the activities, what is very important is the cost of capital and the final rate of return, as well as the relationship between ownership and leverage [17]. Firms with a higher ownership concentration will have access to better conditions when issuing debt, since the blockholders' commitment to the business will be seen as more reliable. But blockholders have to balance the trade-off between the need for funds and the costs associated with a dilution in control [11, 22].
- Investment decisions – which is all decisions connected with building the appropriate portfolio of assets in the company. Due to the limited quantity of funds, it is very important to choose such investment opportunities which bring the company the best profits and thus increase its wealth [14, 26, 27, 29]. This type of decision-making is subcategorized into long–term investment decisions which are connected with fixed and intangible assets (the capital budgeting) and the short–term investment decisions, related to the current assets (the working capital management). The capital budgeting needs to reconcile the cost of invested capital and the future cash flows with the change in money value. Therefore, it is crucial for an assessment of the effectiveness of an investment to choose the proper discount rate, predict the most probable cash flows reached by the investment project and adopt the most suitable payback period. Because all investment projects carry the risk, the effectiveness of the investment should include the premium which is expected for taking

the risk. The investment decisions have to be based on the appropriate method of assessing the effectiveness of the investment project and take into account all aspects of the investment.

The working capital management, on the other hand, relates to the allocation of funds as among cash and equivalents, receivables and inventories [6, 30]. Thus, the short–term financial decision is connected with tradeoff between liquidity and profitability, and according to Gill et al. [8] is one of the most important factors that directly impact the financial performance and shareholders' wealth

• Dividend decisions – meaning all of the decisions connected with the disbursement of profits back to investors who supplied capital to the firm [2, 3, 21]. It could be done by paying the dividends to shareholders or by the share buybacks. Despite the method, the financial decision in the scope of the dividend is concerned with the quota of profits to be distributed among the shareholders. A higher rate of dividend might raise the market value of shares and thus maximize the shareholders' wealth. In the scope of the dividend decisions there are also: dividend stability, stock dividend and cash dividend [7].

The financial decision making is the key management challenge for every firm and the effectiveness of this process is crucial for the growth and survival of the business [25, 36]. On the other hand, the strategic decision making tasks, such as financial decisions, are strongly influenced by an owner – manager's personal behavioral attitudes [12, 35] and their ability to cooperate within the managerial board. Thus, they are mostly the group decisions.

3   Group Decisions and Effectiveness

According to Harrison [10] the group could be understood as: (1) a collective entity – this "being" is independent from the characteristics of its members; (2) a set of individuals – the behaviour of the group is a function of the characteristics of its members; (3) a collective entity composed of a set of individuals – the understanding of the behaviour of the group requires the analysis of characteristics of the group itself as well as its members. The third meaning suggests that the groups have minds of their own and the behaviour of the group could be analysed independently from the characteristics of the group members. This conception is named as a *group mind* and was explored by Wegner [37], introducing a *transactive memory system*. He suggests that transactive memory is a typical group attribute and could not be imputed (is not transable) to a single person/individual as well as found somewhere "between" the group members. It involves the operation of the memory systems of the individuals and the processes of communication that occur within the group. It is rather a property of a group. Having the transactive memory the group should be better prepared to the decision making, as well the effectiveness of this decisions should be higher [20, 28, 34].

There exist many advantages of decision making process within the group. Firstly, the group has the better summarized knowledge and more information than anyone of its members [31, 33]. Thus, if a decision requires use of knowledge, the group should have a better result than individuals. Secondly, the group as a set of individuals could look on the problem from different perspectives, whereas the individuals are usually used to think routinely [18]. The solutions and the decisions made by the group should be therefore much more innovative and uncommon. Another advantage is connected with acceptance of decision. The decisions made by a group have higher level of general acceptance because the membership in the decision process increases the level of responsibility for choice which was made as well as understanding the solution [32].

It should not be forgotten, that the group decision making process has the disadvantages as well [24]. Firstly it is the time needed to the whole process. The group decisions are made much more longer than decisions made by the individuals. Taking consultancies, negotiations etc. is time-consuming. The appearance of several alternatives often causes the members to support a particular position. These preferences take precedence over finding the best solution and the result is a compromise decision of lower quality. In many cases the group accept the first solution which seems to be most suitable for the majority of members or only vocal minority. The other, sometimes better, solutions have lower chance to be taken into account and analysed. Another disadvantage is connected with a strong personality of one or more members of the group comparing to the others. The strong personality might want to impose her or his solution/decision. In such a situation the other members of the group usually do not want to cooperate and engage themselves, so the group cannot rely on the effect of synergy. When the other members want to "deserve" acceptance of the group, they could agree with proposed decisions and tend to behave in accordance with something which is called *groupthinking* – pathological cohesion with the group [19].

Summarizing, there are many arguments which confirm better effectiveness of group decision making as well as the proven opinions and situations which suggest something opposite. Thus, there is no consensus about the correlation between effectiveness and group decision making.

4   Methodology

The research consisted of a simple simulation game done with some students of finance and management, after finishing their course about financial decisions. Thus, the students had more or less the same knowledge and skills in tools supporting the decision making processes in finance.

The students from the whole lecture group were asked either to split into small teams with maximum four members or to enter the game as individual participants. Each of the newly created units had to decide about three things: offered price of the product, size of monthly production and the capital structure (debt ratio) used to finance their activity. The whole lecture group was treated in the game as a market and the decision groups - as companies which make the same goods. Thus, they had to compete with each other about the bids of buying their products. The competition was not purely artificial because the group which obtained the best result got extra points and thereby improved their mark for the whole course.

There were certain limits of the decisions, e.g. the highest offered price per unit could be 290PLN, the highest production – 1500 units and the highest level of debt – 70% of all capital. Each group had the same conditions of production and the same operating costs. Before making the decisions they knew the demand (on the whole market) for the next month, but they did not know about decisions made by other groups.

After gathering the offers, the coach decided how many goods to buy and from whom. The only criterion taken into account was the price. Thus, if one group decided to offer the highest price, it was very possible that nothing would be purchased from them. In such a situation the group would have a loss, because the products have very short term of use.

The main goal of the game was to maximize the profit and the ROE (as the basic indicator of efficiency).

To the analysis was collected information from 540 such games/decisions and they focused on the following variables: group size (g_size) – the integer number from the set <1, 4> , offered amount of production (d_quantity) – the integer number from the set <0, 1500>, offered price (d_price) – the real number from the set <0, 290>, debt ratio (d_debt) – the real number as a percentage value from the set <0, 70> , resulted profit (e_profit) – the real number from the set <−191 400, 245 000> and the resulted ROE (e_ROE) – the real number from the set <−1.6, 2.03>. The extreme values of every variable connected with the decision were limited to the predefined levels, where the values of the variables connected with efficiency were directly derived from the extreme values of decision's variables. The basic characteristics of this data are placed in Table 1.

As could be seen from Table 1 the groups consisted mostly of four people and decided to produce as much as possible (the highest quantity was 1500 units and the average quantity from all analyzed decisions was almost 1235 unit). If the production is on the maximum level (maximum capacity), the lowest price given by the break-even point (BEP) is 126,67. The decrease in production causes an increase in price given by the BEP. In the analyzed group decisions the average offered price was 193,17 and it is a little bit below the middle between the lowest possible price and the highest one. This two figures were completed by the chosen capital structure. The share of debt in whole capital on average was on the level of 36%, what is somewhere in the middle between the lowest possible level (0%) and the highest possible level (70%). Above decisions give the extremely different effects – losses and profits, what has its consequence in the ROE.

Table 1. Descriptive statistics of data

	N	Minimum	Maximum	Mean	Std. deviation	Skewness		Kurtosis	
						Statistic	Std. error	Statistic	Std. error
g_size	540	1,0	4,0	3,51	0,68	−1,177	0,105	0,668	0,210
d_quantity	540	0,0	1500,0	1234,61	297,17	−1,110	0,105	1,143	0,210
d_price	540	0,0	290,0	193,17	46,89	−0,037	0,105	0,333	0,210
d_debt	373	0,0	70,0	36,38	26,15	−0,183	0,126	−1,453	0,252
e_profit	540	−191400,0	245000,0	42102,72	84171,74	−0,718	0,105	0,937	0,210
e_ROE	540	−1,60	2,03	0,22	0,43	0,203	0,105	3,611	0,210
Valid N (listwise)	373								

Because almost each from above data seems to have a normal distribution (except the ROE), to the analysis and verification of the hypothesis could be used the ANOVA, independent and dependent paired samples test.

5   Results

Looking at the statistics divided according to the value of the g_size (Table 2) there could be observed differences in effectiveness between the various categories of the groups as well the differences in certain decisions.

Table 2. Frequencies

g_size			d_quantity	d_price	d_debt	e_profit	e_ROE
1,00	N	Valid	4	4	4	4	4
		Missing	0	0	0	0	0
	Mean		1100,00	227,00	49,00	−43105,00	−0,2479
	Median		1050,00	236,50	51,50	−34130,00	−0,1962
	Std. deviation		294,39	29,54	7,61	134248,67	0,7086
2,00	N	Valid	45	45	45	45	45
		Missing	0	0	15	0	0
	Mean		1298,89	197,91	37,87	47768,44	0,2355
	Median		1400,00	200,00	38,50	65000,00	0,2696
	Std. deviation		286,33	34,90	24,60	79297,59	0,3099
3,00	N	Valid	165	165	96	165	165
		Missing	0	0	69	0	0
	Mean		1195,94	195,33	27,40	30056,27	0,1292
	Median		1200,00	190,00	25,00	37000,00	0,1138
	Std. deviation		296,72	46,85	24,36	95452,72	0,4339
4,00	N	Valid	326	326	243	326	326
		Missing	0	0	83	0	0
	Mean		1246,97	191,01	39,53	48463,26	0,2639
	Median		1350,00	130,00	50,00	49200,00	0,1963
	Std. deviation		297,36	48,40	26,46	76984,46	0,4268

To check the significance of these differences the ANOVA analysis was used. The results could be observed in Tables 3 and 4, below.

Table 3. Analysis of the decision made in regardless with the group size

		Sum of squares	df	Mean square	F	Sig.
d_quantity	Between groups	554950,35	3	184983,45	2,108	0,098
	Within groups	47045239,53	536	87770,97		
	Total	47600189,89	539			
d_price	Between groups	7888,03	3	2629,34	1,197	0,310
	Within groups	1177388,30	536	2196,62		
	Total	1185276,33	539			
d_debt	Between groups	10870,82	3	3623,61	5,490	0,001
	Within groups	243546,88	369	660,02		
	Total	254417,70	372			

Table 4. Analysis the effectiveness of decision made with regardless of the group size

		Sum of squares	df	MeansSquare	F	Sig.
e_profit	Between groups	67619077650,52	3	22539692550,17	3,221	0,022
	Within groups	3751132658190,70	536	6998381824,98		
	Total	3818751735841,21	539			
e_ROE	Between groups	2,870	3	0,957	5,352	0,001
	Within groups	95,808	536	0,179		
	Total	98,678	539			

Analysing the decisions made by the groups with various size, the F statistic suggests that only in the case of debt level the differences are significant. Other decisions, like price and offered production show no differences between groups from the statistical point of view. Thus it seems that size of the group could influence the way of decision making only in the case if the decision is connected with higher risk. It could be supposed, too, that if the decision takes into account the situation or issues which are easier to understand and to predict the results of, the size of the group is not important. When the situation requires to analyse the risk and is not so easy to forecast the effect, the size of the group will be significant.

In order to better understand the differences between the groups, the independent sample tests were used. From these analyses could be seen that decision about the monthly production differs only between the group with 2 members and 3 members. The first one decided on average to produce more than the second group. The mean difference was 102 units, which is 6.8% of the highest possible production. In all other cases it was not significant.

In the decision about the price the groups do not differ noticeably.

The most significant differences could be observed in the decisions about the capital structure. From the statistical point of view the groups with 1 and 3 members, 2 and 3 members, and 3 and 4 members demonstrate marked dissimilarities. The group with 3 members has much lower level of debt than the group with 1 member (the individual). The mean difference is 21% points. The same situation is in the group with 2 and 3 members. The first one decides about higher level of debt than group with 3 members. The mean difference is lower than between the group with 1 and 3 members and amounts to 10% points. The opposite direction of difference could be observed in the pair: group with 3 members and 4 members. The first one has a lower level of debt and the mean difference is about 12% points. The results suggest that the group with 3 members varies significantly from the other groups but the direction of the differences is not homogeneous. When the group has fewer members than three, it is more willing to risk and decides about higher level of debt. When the group has more members than three, it decides about a lower level of debt. It is not compatible with the theorem that risk taking by the group is higher when the group is bigger. This is connected with dilution of the responsibility. Having more members in the group means lower responsibility for the tasks performed. However, the results of the analysis show something different.

The special character of the group with 3 members was also observed in the research made by Laughlin with team [23]. They suggest that the 3 – person groups are necessary and sufficient to perform better than the best individuals. Thus, the decisions made by the 3-person groups should be most effective.

Looking at the effectiveness and the group size, it could be observed (see Table 4) that the profit as well as the ROE show significant statistical differences. According to the results of the independent sample tests, it could be seen that if effectiveness is measured by the profit, only groups with 3 and 4 members expose a noticeable difference. In other cases the variations are statistically insignificant. In this case the profit is higher if the group is bigger. However, the basic statistics (see Table 2) implies that the best results, measured as median value in profit has the group with 2 members, whereas the group with 4 members is on the second position. From the other side, the effectiveness obtained by the group with 1 member (exactly this is an individual, not a group) is below zero and is the lowest of all results, although taking into account the t-statistic this difference is not significant. Thus, it seems that it is not truth that the decision made by a group (not individual) is more effective and the size of the group does not matter.

When the ROE is taken into account, the two pairs have significant differences. The first pair is the group with 1 and 4 members and the second pair is the group with 3 and 4 members. In the first case, the ROE is much lower in the group with 1 member (mean: −24.79%) than in group with 4 members (mean: 26.39%). It is worth noting that the difference is quite big. In the second case, the dissimilarity is smaller (mean difference: 13.47 pp) and every ROE is positive. Of course, it is important to remember that variable e_ROE has not normal distribution and the findings could have more or less mistakes. However, it is observed that the effectiveness measured by the ROE could differ between the individuals and the group, but the size of the group seems have no influence on the results. This could be true because all the groups are small. If we compared small and big or oceans groups, the differences could be much more visible.

The last analysis was connected with the influence of decision making in the case of the failure obtained in the former decision. The question was how the group behaves when the last decision was unsuccessful and whether it is dependent on the size of the group.

Because the coach in the game decided about buying or not of the offered produce at the offered price, the group can experience a loss if they sell nothing or less than they offer. Thus, either decision could be assigned to group (1) – after success or to group (2) – after fail. On such basis the dependent pair samples test was made and according to the t – statistics it was found that the decision about production and price does not change after the fail game phase, but the decision in the scope of capital structure is completely different. What is not surprising, the level of engaged debt is increased after the fail and it is compatible with the psychological effect named the effect of reflection (see [15]). It is a symmetric change in preference against the risk: in the context of making profits one avoids financial risk, but in the situation of making losses one becomes a gambler. It could be supposed that the group behave in the same way and give in the above effect.

6   Summary

Although the individuals make many decisions in both the private and public sector, the truly important decisions are generally assigned to a group [31]. A good example of this are the financial decisions, which are a natural process for companies, made usually within the group. The effectiveness of these decisions builds the wealth of the firm, its managers and owners and should be understood not only in the context of an economic result but also in the context of time and complexity of the whole decision process. The above research shows that financial decision process made by a group is not always more effective than one made by individuals and as the size of the group is increasing it is not true that effectiveness is increasing as well. This finding is compatible with research made by Kerr [19, 20]. They have consistently found that usually groups not only fail to exceed a normative baseline given by the performance of the individual, but also often fall short of it.

Although the findings of the analysis of the game seem to confirm the thesis, known from the earlier research, they have their limitations. Firstly, the amount of information gathered about decision making by an individual is substantially less than that made by a group. Thus, the results of a comparison could be not as precise as they would be with more observations.

Another issue is connected with the research participants. Students very often have no experience in making financial decisions, therefore they may behave differently from business people in a real situation. However, they had a theoretical background which should help them to make those decisions. In the future it will be more advisable to select the participants and gather the data from the target group.

The findings should be even more interesting if they include an analysis of the characteristics of the group members, e.g. experience, attitude to risk, power of influence, etc. The knowledge about the most efficient structure of the group would be very useful for many organisations and institutions.

Finally, the research took into account only the capital structure as one of the financial decisions. It could be interesting to get an insight into the efficiency of group decision making in other financial problems.

References

1. Autukaite, R., Molay, E.: Cash holdings, working capital and firm value: evidence from France. Bankers Markets Investors **132**, 53–62 (2014)
2. Bae, S.C., Chang, K., Kang, E.: Culture, corporate governance, and dividend policy: international evidence. J. Financ. Res. **35**(2), 289–316 (2012)
3. Bancel, F., Bhattacharyya, N., Mittoo, U.R., Baker, H.K.: Cross-country determinants of payout policy: European firms. In: Baker, H.K. (ed.) Dividends and Dividend Policy, Kolb Series in Finance, pp. 71–93. Wiley, Hoboken (2009)
4. Boyett, J.H., Conn, H.P.: Workplace 2000: The Revolution Reshaping the American Business. Dutton, New York (1992)
5. Copeland, T.E., Weston, J.F., Shastri, K.: Financial Theory and Corporate Policy, vol. 3. Addison-Wesley, Boston (1983)
6. Deloof, M.: Does working capital management affect profitability of Belgian firms? J. Bus. Financ. Account. **30**(3–4), 573–588 (2003)
7. Denis, D.J., Osobov, I.: Why do firms pay dividends? International evidence on the determinants of dividend policy. J. Financ. Econ. **89**(1), 62–82 (2008)
8. Gill, A.: Financial policy and the value of family businesses in Canada. Int. J. Entrep. Small Bus. **20**(3), 310–325 (2013)
9. Gruenfeld, D.H., Mannix, E.A., Williams, K.Y., Neal, M.A.: Group composition and decision making: how member familiarity and information distribution affect process and performance. Organ. Hum. Decis. Process. **67**(1), 1–15 (1996)
10. Harrison, E.F.: The Managerial Decision – Making Process. Houghton Mifflin Company, New York (1999)
11. Harvey, C.R., Lins, K.V., Roper, A.H.: The effect of capital structure when expected agency costs are extreme. J. Financ. Econ. **74**(1), 3–30 (2004)
12. Heck, R.K.Z.: A commentary on entrepreneurship in family vs. non-family firms: a resource-based analysis of the effect of organizational culture. Entrep. Theory Pract. **28**(4), 383–389 (2004)
13. Jackson, S.: Team composition in organisations. In: Worchel, S., Wood, W., Simpson, J. (eds.) Group Process and Productivity, pp. 138–176. Sage, London (1992)
14. Jensen, M.C.: Value maximization, stakeholder theory, and the corporate objective function. J. Appl. Corp. Financ. **22**(1), 32–42 (2010)
15. Kahneman, D., Tversky, A.: Choices, values and frames. Am. Psychol. **39**, 341–350 (1984)
16. Katzenbach, J.R., Smith, D.K.: The Wisdom of Teams. Harvard Business School Press, Cambridge (1993)
17. Keasey, K., Martinez, B., Pindado, J.: Young family firms: financing decisions and the willingness to dilute control. J. Corp. Financ. **34**, 47–63 (2015)
18. Kerr, N.L., MacCoun, R.J., Kramer, G.P.: When are N heads better (or worse) than one? Biased judgment in individuals versus groups. In: Witte, E., Davis, J.H. (eds.) Understanding Group Behavior: Consensual Action by Small Groups, vol. 1, pp. 105–136. Lawrence Erlbaum, Mahwah (1996)
19. Kerr, N.L., Tindale, R.S.: Group performance and decision making. Annu. Rev. Psychol. **55**, 623–655 (2004)

20. Kerr, N.L.: The most neglected moderator in group research. Group Process. Intergroup Relations **20**(5), 681–692 (2017). https://doi.org/10.1177/1368430217712050
21. La Porta, R., Lopez-de-Silanes, F., Shleifer, A., Vishny, R.W.: Agency problems and dividend policies around the world. J. Financ. **55**(1), 1–33 (2000)
22. Liu, Q., Tian, G.: Controlling shareholder, expropriations and firm's leverage decision: evidence from Chinese non-tradable share reform. J. Corp. Financ. **18**(4), 782–803 (2012)
23. Laughlin, P.R., Hatch, E.C., Silver, J.S., Boh, L.: Groups perform better than the best individuals on letters-to-numbers problems: effects of group size. J. Pers. Soc. Psychol. **90** (4), 644–651 (2006)
24. Laughlin, P.R.: Group Problem Solving. Princeton University Press, Princeton and Oxford (2011)
25. Mahérault, L.: Is there any specific equity route for small and medium-sized family businesses? The french experience. Fam. Bus. Rev. **17**(3), 221–235 (2004)
26. Modigliani, F., Miller, M.H.: The cost of capital, corporation finance and the theory of investment. Am. Econ. Rev. **53**, 261–297 (1958)
27. Morellec, E., Smith, C.W.: Agency conflicts and risk management. Rev. Financ. **11**(1), 1–23 (2007)
28. Morgan, P.M., Tindale, R.S.: Group vs. individual performance in mixed motive situations: exploring an inconsistency. Organ. Behav. Hum. Decis. Process. **87**, 44–65 (2002)
29. Myers, S.C., Majluf, N.S.: Corporate financing and investment decisions when firms have information that investors do not have. J. Financ. Econ. **13**(2), 187–221 (1984)
30. Raheman, A., Nasr, M.: Working capital management and profitability–case of Pakistani firms. Int. Rev. Bus. Res. Pap. **3**(1), 279–300 (2007)
31. Tindale, R.S., Kameda, T., Hinsz, V.B.: Group decision making. In: Sage Handbook of Social Psychology, pp. 381–403 (2003)
32. Tindale, R.S., Talbot, M., Martinez, R.: Decision making. In: Group Processes, pp. 165–192 (2013)
33. Tyler, T.R., Smith, H.J.: Social justice and social movements. In: Gilbert, D.T., Fiske, S.T., Lindsey, G. (eds.) The Handbook of Social Psychology, 4th ed., vol. 2, pp. 595–629. McGraw Hill, Boston (1998)
34. Thompson, L., Peterson, E., Brodt, S.E.: Team negotiation: an examination of integrative and distributive bargaining. J. Pers. Soc. Pers. **70**, 66–78 (1996)
35. Van Auken, H.E.: A model of small firm capital acquisition decisions. Int. Entrep. Manag. J. **1**(3), 335–352 (2005)
36. Van Auken, H.E., Kaufmann, J., Herrmann, P.: An empirical analysis of the relationship between capital acquisition and bankruptcy laws. J. Small Bus. Manage. **47**(1), 23–37 (2009)
37. Wegner, D.M.: Transactive memory: a contemporary analysis of the group mind. In: Mullen, J.B., Goethals, G.R. (eds.) Theories of Group Behaviour. Springer, New York (1987). https://doi.org/10.1007/978-1-4612-4634-3_9

Diffusion of Electric Vehicles: An Agent-Based Modelling Approach

David M. Ramsey[✉], Anna Kowalska-Pyzalska, and Karolina Bienias

Department of Operation Research, Faculty of Computer Science and Management,
Wroclaw University of Science and Technology, Wroclaw, Poland
david.ramsey@pwr.edu.pl

Abstract. In recent years, most European governments have clearly stated their aims to promote the production and sale of electrical vehicles (EVs), which are seen to be an environmentally-friendly means of transport. At the beginning of 2018, the Polish government introduced legislation to promote the diffusion of electric vehicles. In conjunction with this, the authors present some preliminary work on a major study on the diffusion of EVs on the Polish market. This project aims to simulate the behaviour of the market using an agent-based model, which will be based on the results of a survey carried out among the purchasers of new cars in two major Polish cities, Wrocław and Katowice. Agent-based models allow the decision of agents to be affected by interactions with neighbours. The diffusion of EVs in Poland is at its very beginning, but according to experts the perspectives for further market development are promising. This article describes some of the basic factors behind a household's decision to buy an EV. A simple model is presented, together with a discussion of how such a model will be adapted to take into account the results from the study.

Keywords: Electric vehicles · Diffusion of innovation
Agent-based modelling · Influence of neighbours

1 Introduction

According to the Institute of Citizens' Affairs, flue gas causes about 40% of air pollution in cities, of which 70% is generated in traffic jams [45]. The World Bank indicates that transportation is contributing to the increase in greenhouse gas emissions [46]. At the same time, the demand for new cars is still growing. Data published by the European Automobile Manufacturers' Association indicate that the registration of new passenger cars each month in 2017 was higher than in the corresponding month from 2016, the increases ranging from 2.2% to 11.2% [44]. As one of the responses to the air pollution caused by transportation, scientists, politicians and local authorities have refreshed the idea of electric vehicles (EVs). Such transportation has existed since the middle of the 19th century, but its fast development and diffusion only began recently. EVs

N. T. Nguyen et al. (Eds.): TCCI XXXI, LNCS 11290, pp. 118–135, 2018.
https://doi.org/10.1007/978-3-662-58464-4_11

have gained attention, especially in the U.S. and Europe, due to technological developments and an increased focus on renewable energy sources and energy efficiency. Nowadays, among EVs we can distinguish several categories, such as: fuel cell electric vehicles, hybrid electric vehicles, plug-in electric vehicles and range-extended electric vehicles. Each of these categories is characterised by various advantages and disadvantages and faces different challenges to market growth. However, some of these problems are common to all categories of EVs. At present, the primary problems are a lack of access to refuelling/recharging infrastructure and the costs involved in purchasing and maintaining such a car.

Recently, the adoption and diffusion of EVs has become an important topic of many research studies. Authors have investigated behavioural factors [24, 27, 36, 39], as well as governmental policy and economic scenarios [3, 22, 24, 25, 33], such as e.g. the effect of the petroleum price on the adoption rate. Apart from studies on social acceptance, see e.g. [2, 4, 21, 26], a great number of models have been developed to examine and explore the perspectives and conditions for the further development of EVs [11, 32, 33, 39, 40].

This paper presents a simple agent-based model of the diffusion of electric vehicles developed using the Netlogo platform, which takes into account the perceived difficulty of adoption (e.g. access to recharging), the socio-economic attributes of users (e.g. their income) and social influence. This is initial work in a large scale project to be carried out in Poland that will involve a survey of the purchasers of new cars, which will be used to develop a more sophisticated model.

This is a very current topic in Poland. In January 2018 the Polish government passed an act aimed at promoting the spread of electrical vehicles. The main policies are to lower the tax paid on such vehicles, develop the necessary infrastructure in cities and along major transport corridors and give local governments the right to grant privileged access to EVs. EVs have been exempted from excise duty (up to 3.1% of the value) and firms have been given tax incentives to buy EVs. Municipal governments are encouraged to give EVs greater rights of access than conventional vehicles (e.g. the right to use bus lanes). It is planned that there will be one million EVs in Poland by 2025 [41]. A pilot study ($N = 500$) regarding pro-environmental attitudes and behaviour amongst Poles, conducted in late 2017, revealed that respondents have a great interest in battery (not hybrid) electric cars [19]. The majority of respondents, 71%, independently of the location of their home, household size, education and age claimed that they would prefer to use an EV if it were offered as a replacement vehicle in the case when their car broke down. It is surprising that the respondents seem to prefer EVs to hybrids, due to the lack of recharging facilites. This may be partially explained by respondents not understanding the difference between purely electric vehicles and hybrids (see e.g. [2]). It is possible that individuals are just stating an interest in finding out about EVs. This is connected to the so called behaviour-intention gap, since the respondents are not faced with any real costs when making this declaration, which as well as being an expression of curiousity is seen as being politically correct (see e.g. [5]). This curiousity opens perspectives for the development of EVs. However, it does not mean that financial barriers (the high costs of EVs) or the practical issues related to making an important purchasing decision will not constrain the final purchasing decisions of consumers.

The model presented includes a threshold which describes the relative cost of an EV compared to a conventional vehicle (CV). This threshold reflects both monetary and practical costs, resulting from, e.g., the inconvenience of regular recharging. Hence, when EVs are on average as easy to use as conventional vehicles, this threshold describes the relative price of an EV. The readiness of a household to buy an EV depends on the practicality of an EV for that household, as well as its income and attitudes, which are modulated by their interactions with neighbours. Hence, although individual households make their own decisions, there is a component of group decision making which leads to the final decision.

The purpose of this paper is to present the factors related to purchasing an EV, including the effect of others' opinions on purchasing decisions. A simple agent-based model (ABM) will be presented. It should be noted that presently the authors are conducting a survey among customers of car salons in two major Polish cities, Wrocław and Katowice, to investigate the factors leading to their purchasing decisions. The results from this survey will be used to develop a more realistic ABM for the diffusion of EVs in Poland. The initial results presented here suggest that unless the perceived costs of purchasing an EV are comparable to the costs of purchasing a conventional vehicle, then the demand for EVs will be virtually non-existent.

The remainder of the paper is structured as follows: Sect. 2 briefly describes modelling the diffusion of EVs. In particular, attention is paid to the socio-economic factors included in the models. Section 3 introduces our agent-based model and presents the simulation algorithm. In Sect. 4, the results of our simulation study are presented. Adaptation of this initial model in order to develop a more realistic model is considered in Sect. 5. Finally, Sect. 6 gives a conclusion and discussion of some policy implications.

2   Review of the Literature on Models of EV Diffusion

A wide variety of models are used for examining the diffusion of innovation in the market [1, 8, 14, 21, 28, 35]. Generally, such models can be divided into two groups: bottom-up models (e.g. simulation or optimization models) and top-down models (e.g. input-output models or macroeconomic models), see e.g. [38] for a review. Further attention will be limited to bottom-up models, to which agent-based simulations belong. One of the most important attributes of agent-based computational models (ABMs) lies in the fact that they can capture irrational behaviour, complex social networks and global scale, see [17, 34]. In certain cases, ABMs can be even more appropriate than other modelling approaches, because the latter might face difficulties in detecting and describing direct, functional and analytical relationships between agents and the overall behaviour of the system [18]. In contrast, ABMs allow autonomous, heterogeneous agents to interact according to specific rules and let the macro-level evolve indirectly and bottom-up [12].

Recently, a few reviews have been published regarding the diffusion of EVs. In [1,8] various models regarding different types of EV have been investigated. In [14] only models that explore the diffusion of EVs based on its refuelling infrastructure were taken into account.

Below, the economic and behavioural factors will be elaborated, together with the social interactions influencing the diffusion of EVs, based on the results of these studies.

2.1 Economic Factors

Among economic factors, above all the own-price and cross-price elasticity of demand are crucial, i.e. the absolute price of EVs, their price relative to CVs and running costs. Financial subsidies are also important. In terms of price, researchers usually examine how sensitive consumers are to changes in the prices of EVs or in the relative prices of electric and conventional vehicles. Moreover, the price of petrol or gas, goods which are complementary to conventional cars, is also studied. As Tran et al. [36], indicate, in order to attract early adopters, the prices of EVs should not differ from the price of conventional cars by more than 20%. Recent survey work in the UK suggests that, when purchasing a vehicle, consumers are particularly sensitive to fuel economy and CO_2 emissions due to increased environmental concern [37]. Gnann et al. [14] emphasized that the cost of batteries, oil and EV usage fluctuate by around 25%. According to Shafiei et al. [24,33], the market penetration of EVs depends strongly on policy support from national governments. The market share of EVs is large when gasoline prices are high, EVs are taxed at a lower rate and there exists good access to recharging, see [33].

The study of Figenbaum and Kolbenstvedt [13] found that around 91% of the purchasers of electric cars in Norway already had another car in the household. As Dallinger et al. [9] state, EVs, especially PEVs, have been diffusing faster in the private residential sector than in the commercial one. This is due to access to private garages. The implications of government policies on the diffusion of EVs are also investigated in [3,22,31]. Simpson [30] indicates that household income also has a highly significant effect on this diffusion.

Finally, access to charging stations is found to be crucial in the decision-making process [14,23,37]. Gnann and Ploetz [14] argued that some refuelling infrastructure is needed before the first users can adopt. Non-users are usually more concerned about access to this infrastructure than the early adopters. Moreover, Gnann and Ploetz [14] emphasized that fuel prices for EVs should be lower than for conventional cars, in order to convince consumers to purchase EVs. In the case of plug-in EVs (PEVs), the duration and frequency of refuelling can be much higher than for conventional vehicles. On the other hand, PEVs can be often recharged in domestic garages, directly from the electricity grid. The role of charging stations has also been investigated by Micari et al. [23].

2.2 Behavioural Factors

From a variety of behavioural factors, authors usually take into account consumers' aversion to risk [32,37], together with pro-environmental attitudes and beliefs [32,39]. In many papers [21,24,39] behavioural rules are based on the segmentation of consumers according to Roger's innovation model into: early adopters, early majority, late majority, and laggards [29]. Tran et al. [36] indicate that early adopters are heterogeneous, being motivated by financial benefits, environmental appeal, the lure of new technology, and vehicle reliability. Above all, financial benefits, supported by pro-environmental behaviour, motivate people to adopt EVs. This suggests, as Tran et al. conclude, that EVs should be marketed by appealing to economic benefits combined with pro-environmental behaviour, in order to accelerate the rate of adoption [36].

Many authors also discuss consumers' perceptions and misperceptions of various types of EVs, as well as gaps in consumers' general knowledge and their lack of awareness [2,21,26]. To overcome such a lack of knowledge, it is necessary to disseminate information via all possible communication channels (at points where EVs are sold, through community-led campaigns, where EV owners and users share their experience and educate the public via word-of-mouth or through government-led educational campaigns etc.), see [2].

In [39] a psychologically realistic model is presented, which accounts for the role of emotions in decision-making. The parameters of the model are based on qualitative and quantitative data. The authors assume that agents make decisions by maximising the coherence of their current beliefs and emotions. Simulations suggest that introducing exclusive zones for EVs in cities could accelerate the early-phase diffusion of EVs more effectively than financial incentives alone, see [39].

2.3 Social Influence

Social interactions and the externalities of a social network have been proved to play a significant role in the diffusion of any innovation [16–18,29]. This is also true for EVs [21,27]. The effect of peer influence has been shown to be essential for the diffusion of EVs by many empirical studies, see for example [2,26,27,39]. Also, a large number of models include social influence in the processes of opinion formation and decision-making, see e.g. [6,8,11,32,35,39]. A literature review provided by Pettifor et al. [27] identified three main types of social influence: interpersonal communication (i.e. word-of-mouth, social media and Internet communities), neighbourhood effect (i.e. information gained from observing vehicles demonstrated by others living in close physical proximity) and conformity to social norms. In the case of the neighbourhood effect, the authors indicate that it helps to reduce perceived technological and social uncertainties. The responsiveness to social influence, especially to social norms, depends strongly on cultural norms. More closely-grouped collectivist cultures (e.g. in China) have a higher propensity to imitate the behaviour of others within their social networks compared to more individualistic, status-oriented cultures (e.g.

Western Europe), see [27]. Finally, some authors have shown that diffusion proceeds much faster in a clustered environment of agents, see e.g. [21,35].

In the next section we will introduce an initial approach to modelling the diffusion of EVs. The model presented will be based on some of the findings discussed above. This model will be developed in the future on the basis of the empirical survey of individuals purchasing new cars.

3   The Model

This section describes a simple agent-based model (ABM) used as the basis for simulations. The purpose of this model is to show how different factors can influence the decisions of individual households when such a household decides between buying a conventional car or an EV. Three main points are considered: the relative costs of CVs and EVs, technical issues and the strength of neighbours' opinions. Although there are already models which use AB simulations [7,21,33,39] to explain the diffusion of EVs, this model is focused on opinion formation. The opinion of each agent is re-evaluated at each step of a simulation in such a way that an agent takes a neighbour's opinion into account only if this opinion is not too distant from its own current opinion.

This model combines both a threshold which describes the relative cost of an EV compared to a CV and a component of group decision making. Such interactions with neighbours reflects the reality of Polish consumers, who adapt their opinion to the opinions of neighbours.

3.1   Agents' Attributes

Each agent is characterized by four factors, one of which, opinion, is allowed to vary over time. Let i denote the i-th agent and t denote time (assumed to be discrete). Firstly, an agent has its own initial opinion $x_{i,0}$ about EVs, which is randomly selected from the interval from 0 to 1, where 0 indicates that an agent is maximally prejudiced against electric vehicles and 1 indicates that an agent's opinion about EVs is extremely positive. Next, an agent is described by the effect of the modes of transport already available to a household. This factor, labelled p_i, is important as it is assumed e.g. that a household is more likely to adopt to EV if it already has access to other means of transportation, usually a conventional car. The value of p_i is randomly selected from the interval from 0 to 1, where those without second cars tend to have low values of p_i compared to those with second cars.

The next factor, q_i, is interpreted as a measure of the likelihood that the users of a car travel less than 50 Km in a day. Although the technological developments in batteries for EVs are rapid, it is assumed that a driver who regularly drives more than 50 Km in a day would be less prepared to buy an electric car, because of possible problems related to recharging the battery during the day. The value of q_i is randomly selected from the interval from 0 to 1.

The last characteristic, r_i, addresses the ease with which an agent is able to charge a car. This value is also randomly selected from 0 to 1, where a value close to 1 means that an agent has easy access to recharging points e.g. in a private garage. To sum up, each attribute is drawn randomly from a uniform distribution on the interval (0,1) where values close to 0 imply a very low propensity to purchase an EV and values close to 1 symbolize a high propensity.

It should be noted that the income of a household is implicitly taken into account by this model. Affluent households will often have several cars and garage facilities and hence will correspond to high values of p_i and r_i. The attributes of households in this initial model take into account some of the factors described above (the convenience of using an EV, opinions, affluence). However, it does so in an abstract manner. The results of the survey will be used to analyse the relation between the purchasing decisions of households visiting car salons and the socio-demographic, attitudinal and behavioural variables covered by the survey and construct a more realistic model.

3.2   Opinion Formation

The simple model assumes that each agent interacts with its four neighbours on a two dimensional grid at each step of the simulation. The opinion of a neighbour affects an agent if the difference between their opinions is less than or equal to a defined threshold v (see [21]). Such an approach is motivated by the findings of Hegselmann and Krause [15] and Deffaunt et al. [10], who stated that individuals ignore the opinions of neighbours which are too different from their own.

Let the present opinion of an agent be $x_{i,t}$, the number of this agent's neighbours whose opinion does not differ from the agent's opinion by more than v be $k_{i,t}$ and $y_{i,t}$ be the average value of the opinions of these neighbours. The updated opinion of the agent is given by

$$x_{i,t+1} = \frac{x_{i,t} + k_{i,t} y_{i,t}}{1 + k_{i,t}}. \tag{1}$$

Table 1. Opinion formation: relationship between the maximum distance for influence and the independence of agents

Maximum distance of influence v	0.1	0.2	0.3	0.4
Avg. number of independent agents	4287	1613	615	268
Percentage	42.87%	16.13%	6.15%	2.68%

Hence, at each step of the simulation, the updated opinion of an agent is calculated as the average of its present opinion, $x_{i,t}$, and the opinions of its neighbours which are not too different from the present opinion of the agent. Results from the simulations show that when the maximum distance allowed to affect opinions is 0.3, then around 6–7% of individuals are independent (i.e. do not take the opinion of their neighbours into account). This proportion is around 16% when this maximum distance is 0.2. The proportions of such independent agents for $k \in \{0.1, 0.2, 0.3, 0.4\}$ are presented in Fig. 1. It should also be noted that when the maximum distance of influence is small, then even if an agent changes its opinion, such changes will be very gradual.

Again, this approach is abstract. In reality, a household's "neighbours" will not just be those who live next door, but will include friends and work colleagues. Hence, it will be necessary to take into account more complex graphs describing the social network of agents. Such a model of a network should include, e.g., the distribution of the distance between people's home and their workplace. The evolution of agents' opinions will also depend on how positively purchasers of EVs react to possessing an EV. This is interrelated with how the recharging infrastructure develops. Thus an improved model should take into account how the dynamics of the diffusion of EVs varies according to these factors.

3.3 Relative Cost of EVs Compared to CVs

The final decision of an agent depends strongly on the level of the threshold T. This threshold can be interpreted as the relative cost of an EV compared to a CV. If there are no systematic differences with respect to the running costs or ease of use, then this threshold may be interpreted as the relative price of an EV compared to a CV. The threshold in the simulations takes one of six levels: 1.0; 1.1; 1.2; 1.3; 1.4 and 1.5. If T has value equal to 1.0, then this indicates that the price of an electric car is equal to the price of a comparable car with a combustion engine. If T is equal to 1.5, then we assume that the price of an EV is 1.5 times higher than the price of CV. This difference between the costs is significant in determining the speed of the diffusion of EVs. When T is large, agents who would otherwise readily buy an EV could abandon the idea of buying an EV, because it is too expensive.

Again, this is an abstract way of modelling the relative cost of an EV and this model will be adapted in the light of the results of the authors' survey. Additionally, the relative costs of EVs will not remain fixed over time. Developments in producing batteries and the recharging infrastructure are likely to mean that the perceived costs of an EV relative to a CV will fall over time (given that the government incentives remained fixed). This agrees with research performed by the UBS Evidence Lab in May 2017 [43]. The production cost of EVs seems to be lower than it was initially assumed. Further growth is expected on the demand side of the European market. Hence, UBS researchers conclude that in the coming few years it is likely that the relative price of EVs will fall to a level similar to conventional vehicles.

3.4 Decision Making

When faced with a decision about which type of car to buy, in abstract terms it is assumed that an agent purchases an EV when its relative propensity to buy an EV exceeds the relative cost of adopting an EV. Otherwise, such an agent buys a CV. Firstly, the ease of agent i to adopt an EV is evaluated on the basis of the three following attributes:

- p_i – the effect of the modes of transport already available to a household
- q_i – the probability that the agent travels less than 50 Km in a day
- r_i – the ease with which the agent can charge a car

The ease of adoption, E_i, is given by the average of these randomly selected attributes. The propensity of agent i at time t to buy an electric vehicle is given by $Z_{i,t}$, where $Z_{i,t} = E_i + x_{i,t}$. When an agent decides to buy a new car, this propensity is compared with the threshold T. If this sum is greater than the threshold T, an agent will buy an EV. Otherwise, an agent will decide to buy a conventional car.

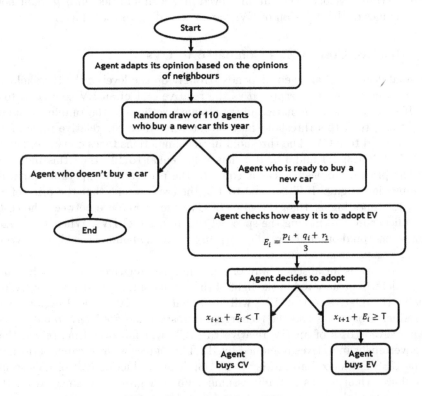

Fig. 1. Steps of decision making.

3.5   Simulations

The simulations are generated in the NetLogo environment. The framework of the model is presented in Fig. 1. In the first step, a two-dimensional network of agents of dimensions 100 by 100 is created. This means that there are 10,000 agents. Each simulation generates the diffusion of EVs over a period of five years, with one step representing a year. In each step of a simulation, after each agent has adapted its opinion regarding EVs as described in Sect. 3.2, 110 agents are selected to buy a new car. This is due to the fact that, according to statistics published by the European Automobile Manufacturers Association in 2016, new 11 cars were sold per 1,000 households. This number includes only private buyers and owners of one-man companies [44]. It should be noted that the model does not include second hand sales, since at present EVs are not available on this market. Also, the number of annual sales of new cars per 1,000 households in urban areas is greater than for the country as a whole, since mean incomes are higher in urban areas. Since government policy will concentrate on developing the recharging infrastructure in urban areas and the surveys are being carried out in major cities, the sales rate given above should be treated as a conservative estimate of the number of sales of new cars for the purposes of the study.

The procedure of deciding whether to buy an EV or a CV is as described in Sect. 3.4. The sum of the present opinion, $x_{i,t}$, and the average value of the agent's other attributes, E_i, is compared with the threshold T.

Initially, five simulations were carried out for each of the 24 combinations of parameters (based on four values for the maximum distance for influencing neighbours v and six for the threshold T). Initially, none of the agents possessed an EV. Analysis of the results indicated that there was very little variation between the simulations based on a given pair of values for these parameters. In addition, the variation across different pairs of parameter values was very systematic. Hence, the results presented below are based on averaging the results from five simulations for each of the pairs of parameter values.

4   Results

This section considers how the sales of EVs over five years depend on the relative cost of EVs and the interaction between neighbours according to the model.

The simulations indicate that the relative cost of EVs has a very clear negative effect on the diffusion of EVs. This finding can be validated by the social study conducted in Poland by the Alternative Fuel Observatory [42] which shows that the price is the most important attribute of EVs for more than 35% of respondents. Our model also indicates that the fastest diffusion of EVs is possible when the costs of conventional cars and EVs are equal (for the same class of a car). However, because of the current costs of the new technologies used in EVs, it seems hard to achieve equal costs in the nearest future without the government offering incentives, see [42].

The results of the simulations indicate that EVs could be successfully introduced onto the market if they were maximally 20% more expensive than conventional cars. In this case, EVs could achieve 20% of the market share in terms of the proportion of the sales of new cars. When the differences in costs are too high (if EVs cost around 40 or 50% more than conventional cars), it would

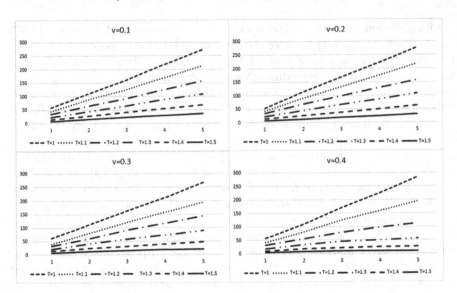

Fig. 2. Number of sales of EVs amongst 10 000 agents over 5 years according to the relative costs of EVs (x-axis: time (in years); y-axis: cumulative number of EVs bought)

be very difficult for EVs to establish any significant market share, which would probably lead to the rejection of the innovation and a lack of investment in the infrastructure needed for EVs to diffuse.

The results also describe how the sensitivity of agents' opinions to the opinions of their neighbours affect the sales of EVs. The parameter v defines what range of neighbours' opinions will affect an agents' opinions. When v is small, agents' opinions do not tend to vary much over time. As v increases, agents' opinions become more similar to the average opinion of their neighbours. Hence, there is a tendency for agents to develop moderate opinions. The results presented in Fig. 2 indicate that the parameter v only has a significant influence on sales of EVs when the threshold T is large (i.e. the relative cost of EVs is large). When EVs are relatively expensive, then for an agent to purchase an EV it must have a very positive attitude towards EVs. Based on the model presented, an agent is only likely to keep such an attitude when v is relatively small.

Table 2 presents the percentage of agents that purchase EVs within 5 years (averaged over five simulations). Each graph corresponds to a maximum distance between opinions that will affect an agent's opinion. The six curves on each graph correspond to the relative costs of EVs, where: $T = 1$ indicates that the costs of comparable EVs and CVs are equal, $T = 1.1$ indicates that EVs are 10% more expensive than CVs, etc. When $T = 1$ (i.e. there is no systematic difference in the cost of EVs and CVs), then approximately 50% of the new cars purchased will be EVs. Some agents will prefer EVs according to their personal preferences, e.g. based on the probability that an agent travels up to 50 Km in a day or ease with which an agent can charge a car. As the cost of EVs relative to CVs

Table 2. Average number of EVs/CVs sold per 100 households after 5 years

	$T = 1$	$T = 1.1$	$T = 1.2$	$T = 1.3$	$T = 1.4$	$T = 1.5$
$v = 0.1$						
CV	2.77	3.37	3.93	4.41	4.82	5.15
EV	2.73	2.13	1.57	1.09	0.68	0.35
st. dev.	0.051	0.058	0.069	0.059	0.043	0.060
$v = 0.2$						
CV	2.72	3.32	3.93	4.42	4.87	5.20
EV	2.78	2.18	1.57	1.08	0.63	0.30
st. dev.	0.156	0.161	0.094	0.071	0.057	0.043
$v = 0.3$						
CV	2.82	3.55	4.07	4.60	5.03	5.29
EV	2.68	1.95	1.43	0.90	0.47	0.21
st. dev.	0.080	0.108	0.082	0.090	0.061	0.031
$v = 0.4$						
CV	2.66	3.56	4.38	4.93	5.23	5.40
EV	2.84	1.94	1.12	0.57	0.27	0.10
st. dev.	0.123	0.076	0.053	0.061	0.044	0.026

increases, the demand for them naturally decreases. This is particularly visible when agents' opinions are not strong (v is large). When $T = 1.5$ and $v = 0.4$, after 5 years on average only one in a thousand agents have purchased an EV (this represents less than 2% of the cars that were purchased in this period).

Regression models were used to verify this interaction between the strength of opinion and the relative cost of EVs. The dependent variable, Y, was the percentage of agents that had bought EVs after 5 years. The independent variables considered were T, v and vT. Here, the coefficients of T and v in a regression model represent the direct effects of the relative cost of EVs and the sensitivity of agents' opinions to their neighbours' opinions. The coefficient of vT represents the effect of the interaction between these two variables. Based on the Akaike information criterion, the model obtained is the following:

$$Y = 7.6422 - 4.8531T - 0.8327vT.$$

This model also maximised the adjusted coefficient of determination. Approximately 96.3% of the variation of sales is explained by this model, which agrees with the qualitative statements made above (price has a negative effect on demand, particularly when agents' opinions are not strong).

5   Future Adaptation of the Model

As stated, the model presented above is a simple, abstract model and should be adapted to the results of a survey that is currently being carried out in car salons in two Polish cities, Wrocław and Katowice. This survey will give information e.g. on the purchasing decisions of those searching for a new car, their attitudes and those of their friends and family, as well as the way in which they travel (e.g. use of public transport/regularity of driving more than 50 Km and the number of cars the household possesses). The characteristics of the agents in the model presented here can be thought of as latent variables that are associated with the likelihood of purchasing an EV. The survey will allow us to analyse the relationship between "the characteristics of real agents" and the likelihood of buying an EV. In the model presented here, the characteristics are assumed to be independent, whereas the survey will be used to analyse the distributions of the characteristics included in the survey, together with the interrelationships between them. It should be noted that exploratory methods of data analysis (e.g. principal component analysis or factor analysis) can be used to define independent latent variables, which can be then interpreted as characteristics of those surveyed and their association with the decision to purchase an EV analysed. As stated previously, the survey will not be representative of the Polish population as a whole, but is concentrated on those who are purchasing a new car and live in or close to a large city. Hence, adaptations of the model should take this into account (e.g. the rate of purchasing new cars will be higher than in the population as a whole).

A more complex model should take into account the fact that agents do not just interact with their neighbours (in the sense of those who physically live "next door" to them), but with friends and work colleagues. This can be modelled by a network where the position of an agent corresponds to the location of a household. Hence, agents will often be linked to their physical neighbours in the network, but will also have links to agents who live some distance away. Also, the number of neighbours/links per agent is not fixed, but will vary according to the household. It is likely that the distribution of the number of neighbours follows a so called scale free distribution. The importance of such a network topology has been emphasized in many scientific papers dealing with the diffusion of innovation and dynamics of opinion formation, see e.g. [6,17,35]. In particular, Sznajd-Weron et al. [35] conclude that the diffusion of innovation is easier on more regular graphs, i.e. those with a higher clustering coefficient. Moreover, the shorter the average path length between agents, the easier it is for an innovation to spread. Such a model should also take into account the location of recharging stations, the network of which will evolve over time. The decision to purchase an EV will depend on the ease with which a car can be recharged (i.e. does the household have its own facilities, are there recharging facilities close to the work place of a member of a household?) and this will change over time.

One assumption of the model presented is that given that on average the perceived cost of an EV is the same as the perceived cost of a CV, then 50% of newly purchased cars will be EVs. There are two obvious problems with this assumption. Firstly, it is difficult to assess precisely what the relative perceived cost of an EV is, due to the constraints imposed by using an EV and the lack of recharging infrastructure at present. Secondly, the novelty of EVs and the distribution of attitudes towards EVs will also determine the initial popularity of EVs on the market. This initial popularity will be investigated by the survey of purchasers of new cars. Many models split consumers into groups, such as early adopters, early majority etc., see [29]. Since the EV market in Poland is still undeveloped, the participants of the survey who buy electrical vehicles can be treated as early adopters and the results of the survey will highlight the characteristics exhibited by this group of consumers.

One obvious difficulty involved in developing a more realistic model lies in the fact that the dynamics of the market for private vehicles will change over time. For example, the population of those buying a new car each year will change depending on the economic situation in Poland. Also, the opinion of those who have bought an EV will begin to play an important role in how the market evolves. At present, EVs are treated as a novelty in Poland and so the decision of whether to buy an EV generally does not depend on concrete information about the day to day practicalities of possessing an EV, but on people's general attitude to new products and EVs, in particular. The decisions of consumers about whether to buy an EV will evolve over time and will depend on how the recharging infrastructure evolves over time. Since it is difficult, if not impossible, to predict how attitudes to EVs and the infrastructure will change over time, it will be necessary to consider various scenarios. It should be noted that the NetLogo environment is flexible enough to develop such models.

6 Discussion and Conclusions

This paper has presented initial work on a large scale study on the introduction and diffusion of EVs in Poland. There are some obvious shortcomings of this research, which the authors plan to address in future work as described in Sect. 5.

Firstly, although the factors considered here do affect households' propensity to buy an EV, the importance of these factors was chosen arbitrarily. In conjunction with car dealers in both Wrocław and Katowice, the authors are presently running a study of households purchasing a new car. This study will allow the authors to estimate the effect of various factors on the propensity to buy an EV. In addition, the simulations presented here assume that the factors considered are independent. In practice, this is not the case. For example, possession of more than one car by a household is positively correlated with having a garage, i.e. easy access to recharging.

Secondly, the grid pattern of interactions considered is simplistic. Future models will consider more natural patterns of social interaction based on the concept of scale free distributions. Thirdly, such a model does not consider the possibility of systematic changes in the opinions of households. In practice, this will depend on how satisfied the purchasers of EVs are with their new cars and how the infrastructure surrounding electrical cars develops. Since these are difficult factors to predict, the authors propose to look at various scenarios. The rate at which new cars are bought will also be an important factor. Recently, this rate has been increasing.

This article was submitted shortly after the Polish government passed legislature on developing the infrastructure for recharging EVs and tax exemptions on their sales. In the future, we also plan to include the influence of the government's policies into the model.

The theory and results presented here give an idea of how quickly EVs could initially diffuse on the Polish market. As the results and previous studies indicate, the most important factor that influences the diffusion is the relative cost of EVs compared to CVs. In simple terms, if the cost of an EV is relatively high, then a household will not buy one, even if the household has a positive opinion about EVs. The results suggest that EVs should be at most 20% more expensive than CVs to attain any significant share of the market. If the relative cost of EVs is higher, the vast majority of Polish consumers will not buy EVs. The model also emphasizes that when the relative cost of EVs is high, the diffusion is sensitive to how strong the views of households are. When the relative cost of EVs is high, then the number of EVs sold will be significant only if households maintain a very high opinion of them (and even then the number of EVs sold increases very slowly).

Due to the difficulty of running EVs in rural areas and the relative wealth of urban areas, the decision to initially concentrate the development of infrastructure in urban areas and the main transport corridors between them seems natural.

If the relative price of EVs cannot be decreased solely on the basis of market interactions between demand and supply, other solutions should be proposed to decrease the costs incurred by consumers. To enhance the diffusion of EVs, the government or local authorities could decrease the value added tax or propose governmental grants and subsidies. Such practices have already been introduced in other European countries with success [20]. One can expect positive results from such incentives, since initial surveys have shown that consumers are positively inclined towards electric cars, see [19]. However, even if the prices of CVs and EVs are comparable, the authors expect that, at least initially, consumers would feel on average that the costs of running EVs are greater than the costs of CVs. This is due to the lack of infrastructure at present.

To summarize, the results of surveys will be used to develop a more sophisticated model of the dissemination of EVs on the Polish market. This model should be representative of car purchasers in major cities at the present time. However, the speed of diffusion of EVs onto the Polish market will depend on the reaction of early adopters to the day to day practicalities of driving an EV and the effectiveness of the new measures introduced by the Polish government. How the market will evolve is uncertain. Hence, the model should consider various scenarios. The diffusion of EVs in more rural areas will be naturally slower.

Acknowledgement. This work was supported by the Ministry of Science and Higher Education (MNiSW, Poland) core funding for statutory R&D activities.

References

1. Al-Alawi, B.M., Bradley, T.H.: Review of hybrid, plug-in hybrid and electric vehicle market modeling studies. Renew. Sustain. Energy Rev. **21**, 190–203 (2013)
2. Axsen, J., Langman, B., Goldberg, S.: Confusion of innovations: mainstream consumer perceptions and misperceptions of electric-drive vehicles and charging programs in Canada. Energy Res. Soc. Sci. **27**, 163–173 (2017)
3. Axsen, J., Wolinetz, M.: How policy can build the plug-in electric vehicle market: Insights from the respondent-based preference and constraints (REPAC) model. Technol. Forecast. Soc. Change **117**, 238–250 (2017)
4. Axsen, J., TyreeHaheman, J., Lentz, A.: Lifestyle practices and pro-environmental technology. Ecol. Econ. **82**, 64–74 (2012)
5. Bond, T.G., Fox, C.M.: Applying the rasch model: fundamental measurement in the human sciences. Routledge, Oxford (2001)
6. Chmiel, A., Sznajd-Weron, K.: Phase transitions in the q-voter model with noise on a duplex clique. Phys. Rev. E **92**, 052812 (2015)
7. Cho, Y., Blommestein, K.V.: Investigating the adoption of electric vehicles using agent-based model. In: Portland International Conference on Management of Engineering and Technology (PICMET), pp. 2337–2345, September 2015
8. Daina, N., Sivakumar, A., Polak, J.W.: Modelling electric vehicles use: a survey on the methods. Renew. Sustain. Energy Rev. **68**, 447–460 (2017)
9. Dallinger, D., Schubert, G., Wietschel, M.: Integration of intermittent renewable power supply using grid connected vehicles - a 2030 case study for California and Germany. Appl. Energy **104**, 666–682 (2013)
10. Deffuant, G., Neau, D., Amblard, F., Weisbuch, G.: Mixing beliefs among interacting agents. Adv. Complex Syst. **3**, 87–98 (2001)
11. Eppstein, M.J., Grover, D.K., Marshall, J.S., Rizzo, D.M.: An agent-based model to study market penetration of plug-in hybrid electric vehicles. Energy Policy **39**, 3789–3802 (2011)
12. Epstein J.M.: Modelling to contain pandemics. Nature **406**(9) (2009)
13. Figenbaum, E., Kolbenstvedt, M.: Electromobility in Norway - experiences and opportunities with Electric vehicles. Institute of Transport Economics, Oslo (2013). https://www.toi.no/publications/electromobility-in-norway-experiences-and-opportunities-with-electric-vehicles-article32104-29.html. Accessed 12 Dec 2017
14. Gnann, T., Ploetz, P.: A review of combined models for market diffusion of alternative fuel vehicles and their refueling infrastructure. Renew. Sustain. Energy Rev. **47**, 783–793 (2015)
15. Hegselmann, R., Krause, U.: Opinion dynamics and bounded confidence. models, analysis and simulation. J. Artif. Soc. Soc. Simul. **5**(3), 1–33 (2002)
16. Karakaya, E., Hidalgo, A., Nuur, C.: Diffusion of eco-innovations: a review. Renew. Sustain. Energy Rev. **33**, 1935–1943 (2014)
17. Kiesling, E., Guenther, M., Stummer, C., Wakolbinger, L.M.: Agent-based simulation of innovation diffusion: a review. Central Eur. J. Oper. Res. **20**(2), 183–230 (2012)
18. Kowalska-Pyzalska, A.: What makes consumers adopt to innovative energy services in the energy market? a review of incentives and barriers. Renew. Sustain. Energy Rev. **82**(3), 3570–3581 (2018)
19. Kowalska-Pyzalska, A.: Determinants of green electricity adoption among residential consumers in Poland: an empirical analysis. In: Proceedings of 15th International Conference on the European Energy Market, Lodz, Poland (2018, in review)

20. Levay, P.Z., Drossinos, Y., Thiel, C.: The effect of fiscal incentives on market penetration of electric vehicles: a pairwise comparison of total cost of ownership. Energy Policy **105**, 524–533 (2017)
21. McCoy, D., Lyons, S.: Consumer preferences and the influence of networks in electric vehicle diffusion: an agent-based microsimulation in ireland. Energy Res. Soc. Sci. **3**, 89–101 (2014)
22. Melton, N., Axsen, J., Goldberg, S.: Evaluating plug-in electric vehicle policies in the context of long-term greenhouse gas reduction goals: comparing 10 Canadian provinces using the "PEV policy report card". Energy Policy **107**, 381–393 (2017)
23. Micari, S., Napoli, G., Antonucci, V., Andaloro, L.: Electric vehicles charging stations network - a preliminary evaluation about Italian highways. In: 2014 IEEE International Electric Vehicle Conference, IEVC (2014)
24. Ploetz, P., Gnann, T., Wietschel, M.: Modelling market diffusion of electric vehicles with real world driving data - Part I: model structure and validation. Ecol. Econ. **107**, 411–421 (2014)
25. Pfahl, S., Jochem, P., Fichtner, W.: When will electric vehicles capture the German market? and why? World Electric Vehicle Symposium and Exhibition EVS (2014)
26. Peters, A., Duetschke, E.: How do consumers perceive electric vehicles? a comparison of German consumer groups. J. Environ. Policy Plann. **16**(3), 359–377 (2014)
27. Pettifor, H., Wilson, C., Axsen, J., Abrahamse, W., Anable, J.: Social influence in the global diffusion of alternative fuel vehicles - a meta-analysis. J. Transp. Geogr. **62**, 247–261 (2017)
28. Ringler, P., Keles, D., Fichtner, W.: Agent-based modeling and simulation of smart electricity grids and markets - a literature review. Renew. Sustain. Energy Rev. **57**, 205–2015 (2016)
29. Rogers, E.M.: Diffusion of Innovations, 5th edn. Free Press, New York (2003)
30. Simpson, A.: Environmental attributes of electric vehicles in Australia (CUSP Discussion Paper. Curtin University Sustainable Policy Institute, Perth, Australia. www.sustainability.com/au/renewabletransport (2009)
31. Sopha, B.M., Kloeckner, C.A., Hertiwch, E.G.: Exploring policy options for a transaction to sustainable heating system diffusion using agent-based simulation. Energy Policy **39**, 2722–2729 (2011)
32. Sopha, B.M., Kloeckner, C.A., Febrianti, D.: Using agent-based modeling to explore policy options supporting adoption of natural vehicles in Indonesia. J. Environ. Policy **52**, 1–16 (2016)
33. Shafiei, E., Thorkelsson, H., Ásgeirsson, E.I., Raberto, M., Stefansson, H.: An agent-based modeling approach to predict the evolution of market share of electric vehicles: a case study from Iceland. Technol. Forecast. Soc. Change **79**(9), 1638–1653 (2012)
34. Stummer, C., Kiesling, E., Guenther, M., Vetschera, R.: Innovation diffusion of repeat purchase products in a competitive market: an agent-based simulation approach. Eur. J. Oper. Res. **245**, 157–167 (2015)
35. Sznajd-Weron, K., Szwabiński, J., Weron, R., Weron, T.: Rewiring the network. what helps an innovation to diffuse? J. Stat. Mech. **3**, P03007 (2014)
36. Tran, M., Banister, D., Bishop, J.D.K., McCulloch, M.D.: Simulating early adoption of alternative fuel vehicles for sustainability. Technol. Forecast. Soc. Change **80**(5), 865–875 (2013)
37. Tran, M., Brand, C., Banister, D.: Modelling diffusion feedbacks between technology performance, cost and consumer behaviour for future energy-transport systems. J. Power Sources **251**, 130–136 (2014)

38. Ventosa, M., Baillo, A., Ramos, A., Rivier, M.: Electricity market modeling trends. Energy Policy **33**(7), 897–913 (2005)
39. Wolf, I., Schröder, T., Neumann, J., de Haan, G.: Changing minds about electric cars: an empirically grounded agent-based modeling approach. Technol. Forecast. Soc. Change **94**, 269–285 (2015)
40. Zhang, T., Gensler, S., Garcia, R.: A study of the diffusion of alternative fuel vehicles: an agent-based modeling approach. J. Prov. Innov. Manag. **28**, 152–168 (2011)
41. Business Insider Polska (in Polish). https://businessinsider.com.pl/technologie/nowe-technologie/ustawa-o-elektromobilnosci-przyjeta-przez-rzad/fh77nd9. Accessed 12 Jan 2018
42. Alternative Fuel Observatory (in Polish). http://www.orpa.pl/temat/badania/. Accessed 16 Jan 2018
43. UBS Evidence Lab, Global Research: UBS Evidence Lab Electric Car Teardown - Disruption Ahead? Q-Series, 18 May 2017. Accessed 16 Jan 2018
44. European Automobile Manufacturers Association. http://www.acea.be/statistics/tag/category/passenger-cars-registrations. Accessed 16 Nov 2017
45. Institute of Citizens Affairs (in Polish). https://inspro.org.pl/tag/lepszytransport/. Accessed 16 Nov 2017
46. The World Bank. http://documents.worldbank.org/curated/en/23804146798627 5608/Reducing-Greenhouse-Gases-GHG-analysis-in-transport. Accessed 16 Nov 2017

Decision Processes Based on IoT Data
for Sustainable Smart Cities

Cezary Orlowski[1]($\boxtimes$), Arkadiusz Sarzyński[2], Kostas Karatzas[3],
and Nikos Katsifarakis[3]

[1] Institute of Management and Finance,
WSB University in Gdańsk, Gdańsk, Poland
corlowski@wsb.gda.pl
[2] Faculty of Management and Economics, Department of Applied Business
Informatics, Gdańsk University of Technology, Gdańsk, Poland
[3] Department of Mechanical Engineering, Environmental Informatics Research
Group, Aristotle University, Thessaloniki, Greece

Abstract. The work presents the using of decision trees for improvement of building business models for IoT (Internet of Things). During the construction of the method, the importance of decision trees and business models for decisions making was presented. The method of using SaaS (Software as a Service) technology and IoT has been proposed. The method has been verified by applying to building business models for the use of IoT nodes to measure air quality for Smart Cities.

Keywords: Internet of Things · Business models · Decision trees
Air quality measurement

1 Introduction

The emergence of Internet of Things (IoT) technologies supports the smart cities approach in traditional urban environmental management. This calls for new business models that will effectively deal with IoT related solutions which are developed in order to serve sustainability priorities. In the frame of this study we deal with urban air pollution as the environmental problem under consideration and we suggest the use of decision trees for the development of related business models taking into account IoT technologies.

The paper is divided into four main parts. Decision-making processes in cities including Smart Cities are discussed in the first part. The second part presents the importance of decision trees and business models for making decisions. The third-technological part focuses on existing data processing technologies for decision trees and business models. It includes Software as a Service (SaaS) and Internet of Things (IoT). The fourth part is the presentation of the method of using decision trees for the construction of a business model that can be applied to the construction of particulate matter (PM10) measurement sensors for IoT nodes which will be used for smart city applications.

© Springer-Verlag GmbH Germany, part of Springer Nature 2018
N. T. Nguyen et al. (Eds.): TCCI XXXI, LNCS 11290, pp. 136–146, 2018.
https://doi.org/10.1007/978-3-662-58464-4_12

2   The Decision-Making Problems of Smart Cities

The complexity of decisions for environmental management departments in Cities has been pointed out [10]. Such an approach is a consequence of the situation in which most of the decisions made aim at seeking optimal solutions. They must take into account many optimization criteria. In this case, we assume that the decision-making process includes a logically connected group of ordered thought operations, enabling the assessment of the decision-making situation and selection of the best variant. The problematic situation is also a factor triggering the decision-making processes.

A Smart City (SC) is a city in which stakeholders (residents and decision makers) solve their problems with the use of information and communication technologies (ICT) towards resource use optimization [1]. Such a city enables the combination of various urban systems and stimulates innovations that facilitate the implementation of policy objectives towards measurable goals like energy saving or environmental quality improvement, thus supporting sustainability. Thus, a Smart City becomes a creative, sustainable city, where quality of life is improved, the quality of the environment becomes better and the prospects of economic development are stronger [2]. The distinguishing feature of such cities is integration, which can be understood as the synergies emerging from various improvements regarding the functioning of the city's urban infrastructure and the use of resources, as well as public services.

Making decisions in Smart Cities then supports the process of acquiring for new residents, investors and tourists. We can compete if we have knowledge about the state of the city and decision-making processes and consciously use it. This knowledge includes information as a consequence of conducting advanced analysis of data obtained from the city's facilities [12]. This knowledge and data is also a consequence of the city's IT platform facilitating cooperation between authorities by directing problems directly to the appropriate units [11].

An example of such activities (having a platform consciously using knowledge resources) is the Danish city of Aarhus. Working groups have been set up in this city that analyse and suggest solutions for problems such as urban transport, health and sustainable growth. As a back bone of these activities, a free Wi-Fi network for communication with residents (providing information) was also launched. The concept of open data has been materialized via a portal with data under the "Open Data Aarhus" program. All service providers interested in the data have access to it.

3   Decision Trees and Business Models

The second aspect of decision making towards sustainable smart cities is the formalization of the decision-making processes. Also discussed in this part of the article are business models being the resources of decision-making processes. It is difficult to make decisions without being aware of the consequences of these decisions. Therefore, it was assumed that the combination of decision trees and business models will create conditions for an attempt to formalize decisions.

The term decision tree is understood as a graph without cycles (loops) in which there is only one path between two different nodes. It represents a process of dividing a set of objects into homogeneous classes. Its internal nodes describe the method of making this division, and the leaf corresponds to classes integrating objects. In addition, the edges of the tree represent the values of the features for which the division was made. It allows to weigh possible actions against one another.

A business model is defined as a plan implemented by a financial institution to generate revenues and thus lead to profits [3]. It turns out that traditional business models are not enough for Smart Cites technology. It is necessary to create new models and hence the search for innovations in the construction of a business model and the search for "New paths to competitive advantage" [4]. Bucherer et al. (2012) [5] described that the design of business models for Smart Cities technology requires the exchange of information and knowledge among stakeholders (municipal office, citizens, public interest groups, business owners, service providers).

In addition, Westerlund et al. (2014) [6] pointed out three contemporary challenges for Smart Cities technology, including: the diversity of facilities, the immaturity of innovation as well as non-structural ecosystems. The concept of diversity of objects refers to the possibility of connecting to the network many different types of objects and devices without commonly accepted or emerging standards. It is also assumed that in Smart Cities technologies, the innovative process is also still immature. In the case of these technologies, ecosystems are also created that have no defined roles for stakeholders or rules that create value chains.

When constructing these models, there are also questions about companies/stakeholders for which business models are constructed. Questions about which service company is right for the model being developed? What is the best organization dealing with development, sales and provision of services - at regional and international level. How can I convince customers to pay for services previously received free of charge. What is the pricing determined for these services? How to organize and motivate my sales organization? Therefore, it is necessary to compare the strategic and operational characteristics of products and services with each other and to develop relations between them for the construction of a business model.

The aim of this article is to use decision trees and business models in Smart Cities. We assume that the concept of building these models is based on a preliminary analysis of the role that the Internet has already played in the creation of business models, data for the needs of these models and the logic of the use of the proposed product/service. It serves as a basis for building specific elements and patterns of Internet of Things business models as well as historical data on the construction of such models. It is also assumed in the construction of these models that the features/components of Smart Cities technology can be integrated and then form the basis for the construction of such models. Then the key challenge in implementing these models is to create an optimal combination of products and services.

4 SaaS Solution for Air Quality Measurements

This part of the paper describes the use of the example Smart Cites technology which is a system for measuring, analyzing and forecasting pollutants (PM10) for the city of Gdańsk, Poland [7]. A feature of this technology is also the integration of services used by Smart Cities. It was also shown how the integration of individual services requires the analysis of decision-making processes and the use of decision trees and business models in this analysis. This approach is used in the construction of large information systems that use powerful data resources to make decisions. The following processes are also necessary: installation of the IT production environment, connection of the database, development of flow mechanisms and data presentation.

For this reason, before starting the software development a decision was made to choose the architecture of the future system. The system based on SOA (Service Oriented Architecture) supported by ESB (Enterprise Service Bus). Subsequently, the requirements for the system were defined, the data architecture was designed based on the Microsoft environment and applications based on RAS (Rational Software Architect). The integration bus was built based on the Broker Toolkit. The IT system is illustrated based on the IOC (Intelligent Operating System). It consists of three layers, forming an integration bus that provides data flow. The data layer (Database Input node) creates the conditions for connecting to the integration database bus (in our case, the data warehouse akwilon2). The processing layer (node Mapping node) converts data from the data warehouse to the protocol of the CAP (Common Alerting Protocol). In addition, the MQ Output Node node is designed to place events in the CAP format in the integration bus queue manager. The IBM WebSphere Message Broker Toolkit (bus structure) and Netcool Impact (placing events on the IOC map) were the environment for the implementation of the system architecture.

Data from 8 stations of the municipal office were selected for analysis. One of the sensors was placed at the PM10 sensor of the air quality foundation (Armaag Foundation) in order to compare the results with those received from stations belonging to this foundation (treated as reference). The second sensor was located at the Gdańsk University of Technology. The choice of this location was a consequence of extending the measurement network with measurement stations belonging to the Faculty of Civil and Land Construction of the Gdańsk University of Technology. The remaining 6 sensors were located at the City Hall's measurement stations. These stations were equipped with traffic radars measuring traffic volume. The placement of sensors at these stations is the necessity of current and subsequent analysis of the impact of traffic intensity on the level of PM10 in the Gdańsk agglomeration [8]. The sensors have been calibrated and installed at the noise stations.

In the case of data from the Gdańsk University of Technology, a similar solution was used as in the City Hall. The server and its resources were designed, and then the bus power supply for IOC was started on-line. In addition, integration tests (to existing data) of noise levels were carried out. It was assumed that the map acquisition process for the ESB bus will have a two-step character: in the first one, maps in the form of batch files will be delivered, in the second they will supply the system via the access mechanisms built-in using the Message Broker Toolkit.

In addition to data testing and scenario analysis, IOC and ESB have been integrated. The possibilities of building decision trees based on the data attached to the bus were also examined.

5 IoT Nodes - Solution for Air Quality Measurements

The third chapter describes the use of SaaS-based technologies for Smart Cities. We are currently focusing on the description of IoT technology that can be integrated with SaaS technologies. Data from IoT nodes are processed using the ESB integration bus. At the beginning, the IoT network will be characterized. Later, the application of this network for air quality measurements for Smart Cities will be shown.

When writing about IoT measurement networks, we should start with the definition of IoT. Internet of Things is the combination of devices centred around sensors, actuators, presence and positioning, distributed computer power, wired and wireless communication on the hardware side, and data collection/warehousing, applications and data analytics on the software side. It constitutes a major enabler for the use of data mining, and, predictive analytics and other big data techniques (Modified, derived from a Morgan Stanley Research definition).

The current development of the Internet of Things is correlated with the development of networks of measuring devices that can be connected to this network, which does not require high data rates. The official forecast for 2022 is 29 billion devices, including 18 billion operating in IoT (Hans Vestberg, Ericsson, 2010).

The example of IoT applications are measurement networks. Examples of such networks are measurement networks for measuring PM using IoT nodes from companies such as Intel, Bosch, AQMesh, and Yuktix. Their construction is a consequence of projects carried out by these companies aimed at building monitoring stations. The construction of these stations is essential to monitor pollution and problems arising from these measurements. Therefore, a wide survey is being carried out related to the assessment of the usefulness of simple monitoring stations, reference stations, combining them with each other and assessing the quality of the acquired data. Available literature sources show how important it is to use in these tests sensor stations and quality evaluation of dusts with fractions PM10, PM2.5 obtained from these data stations [9].

The article focuses on describing the solution of the Bosch company which is based on the Intel processors. The processor series called Quark is designed for IoT. The advantages of this type of construction are: small size and integrated digital transducer, which provides the possibility of connecting multiple sensors at the same time. Other advantages of the solution are the built-in Bluetooth LE, an accelerometer, or even a controller enabling the battery hold. This system provides the necessary functionalities. At the same time, this system maintains a very low demand for electricity. The main functionalities of the IoT node are:

- Measurements of basic types of air pollutants: particulate matter, carbon monoxide, nitrogen oxide, nitrogen dioxide, sulfur dioxide, ozone as well as environmental parameters, i.e. temperature, relative air humidity, light, sound pressure and air pressure.
- Plug-n-play interface for sensors, so you can easily connect them. They are also recognized by the base station.
- High level of security provided by multi-level Quark processor security.

The main benefits of this solution are advanced analytics and data management enabling:

- Preliminary classification of extreme data and rejection, guarantee of real-time data readout, accurate data collection on a number of important microclimate parameters
- Forecasting trends at the time and place of occurrence - based on benchmarks for key environmental factors such as air pollution, temperature, humidity, etc.
- Standardized graphical interface available on many platforms. Real-time alerts and notifications in nearby regions are available on devices based on various platforms such as tablets and smartphones.

The authors of the work proposed for the development of the Internet of Things network in Gdańsk the construction of a measurement system using IoT nodes for measuring PM10. The system meets the following requirements: platform popularity, a large number of available libraries, easy connection of sensors, wireless communication capability and low power demand. It uses the Arduino platform, the UNO R3 model as the simplest base for building a sensor station.

Based on this platform, further components have been selected. The most important of them is the Sharp GP2Y1010AU0F dust sensor that is able to measure atmospheric aerosols with particle sizes below 2.5 μm. The sensor itself is not enough for the correct measurement, because the values are strongly dependent on temperature, humidity, or wind strength and direction. A DHT22 sensor - a digital sensor was used to measure temperature and humidity. The wind force and direction test in this project has been omitted. Knowing exactly where the sensors are located, you can download this information from external sources.

Additionally, information was collected regarding: date and time of readings. This operation was aimed at checking whether data is sent on a regular basis and whether there are any delays in data transmission. It was also verified in this way whether there is chaos in the database in case of uneven data generation by a larger number of sensors. A communication module was also selected. The Wifi ESP8266 module was selected due to the low power demand, a simple communication method between the Arduino UNO platform, ease of control and small dimensions.

6   The Use of Decision Trees to Build Business Models

One of many methods supporting the design of decision-making processes is the construction of decision trees. These trees can also be used to build business models for supporting cooperation with experts in the construction of these models. The classic process of building a decision tree includes the construction of the apex and the construction of nodes at any decision level. The process also includes defining leaves to be assigned to the appropriate decision class. When creating a decision tree from N input variables included in any vector x, the decision threshold value is assumed by minimizing the decay decision distribution. The CART algorithm and the Gini measure were used in the construction of the decision tree. A feature of this algorithm is excessive tree growth and trimming individual branches to reduce the description of the leaves. However, CART trees are strictly binary. The construction process involves conducting for each decision-making node an in-depth search of all available variables and all possible divisions. Then the optimal division is selected according to the criterion:

$$\varphi(s|t) = 2P_L P_P \sum_{j=1}^{\#classes} |P(j|t_L) - P(j|t_P)| \tag{1}$$

t_L — left child of the tree t

t_P — right child of the tree t

$$P_L = \frac{the\ number\ of\ records\ in\ T_L}{the\ number\ of\ records\ in\ training\ set}$$

$$P_P = \frac{the\ number\ of\ records\ in\ T_P}{the\ number\ of\ records\ in\ training\ set}$$

$$P(j|t_L) = \frac{the\ number\ of\ records\ included\ in\ class\ j\ in\ t_L}{the\ number\ of\ records\ in\ t}$$

$$P(j|t_P) = \frac{the\ number\ of\ records\ included\ in\ class\ j\ in\ t_P}{the\ number\ of\ records\ in\ t}$$

A decision tree structure was developed to classify the PM10 dust level, taking into account a predefined set of data classes. For the construction of the tree, PM10 data from the Armaag foundation was used from 10 measuring stations.

The following algorithm was used:

- The value of the decision attribute was specified: dust level PM10. Values (less than 10, between 10–20, above 20)
- The values of other attributes were determined: temperature, humidity, wind force
- A set of data has been divided into a training set and test set, in 70/30 proportions
- The node division method has been selected - the CART algorithm

The authors built different models of decision trees, using Rapid Miner technology they created and tested a tree with a higher level of complexity. The commonly accepted methodology was used for preprocessing data in which after obtaining data, it was purified, transformed, created a model and then the predictive properties of the model were tested. As part of the data cleaning process:

- lines with missing attributes have been deleted
- rows with attribute values out of range have been deleted
- the sheet has been cleaned from standard attributes

The results obtained are presented below. Two variables were selected for the first learning of the model - temperature and time of day (D and AA). The possible dividing lines for variables have been identified

- the time of day takes 3 values (dark, gray, light)
- the temperature is in the range of closed <–2.10> Celsius degrees and the measurements are presented to the nearest degree
- the division is possible at one degree Celsius for temperature or yes/no for each time of day

Then, calculations were made, aimed at selecting the first division into the branches of the decision tree. The purpose of these calculations was to divide the heavily weighted Gini Impurity criterion of divided branches by reducing them from the current state (0.3367). Gini Impurity for a single branch is expressed by the formula (2). Next, the value of the coefficient for all records combined - current is calculated. Subsequently, for each possible binary division the value of the ratio of the resulting branches was calculated:

- the coefficient weighted by the number of left and right branches was calculated
- the candidate for the first division was selected

Table 1 presents data for Gini Impurity. As a criterion for the division, the comparison was determined by the limit value of temperature - 3 °C. Such a division will result in a reduction of the weighted Gini Impurity to 0.2376.

Table 1. Gini Impurity method results.

Attribute	Candi-date	PL	PR	PPL	PPR	PPL prz	PPL nieprz	PPR prz	PPR nieprznie	Gini Impurity L	Gini Impurity R	Gini Impurity Weighted
TimeOfDay	DARK	39	31	0,5571	0,4429	0,2564	0,7436	0,1613	0,8387	0,3813	0,2706	0,332270
TimeOfDay	GRAY	13	57	0,1857	0,8143	0,2308	0,7692	0,2105	0,7895	0,3550	0,3324	0,336611
TimeOfDay	LIGHT	18	52	0,2571	0,7429	0,1111	0,8889	0,2500	0,7500	0,1975	0,3750	0,329365
temperature	<-1	5	65	0,0714	0,9286	0,2000	0,8000	0,2154	0,7846	0,3200	0,3380	0,336703
temperature	<0	9	61	0,1286	0,8714	0,4444	0,5556	0,1803	0,8197	0,4938	0,2956	0,321103
temperature	<1	10	60	0,1429	0,8571	0,5000	0,5000	0,1667	0,8333	0,5000	0,2778	0,309524
temperature	<2	16	54	0,2286	0,7714	0,5625	0,4375	0,1111	0,8889	0,4922	0,1975	0,264881
temperature	<3	19	51	0,2714	0,7286	0,5789	0,4211	0,0784	0,9216	0,4875	0,1446	0,237653
temperature	<4	23	47	0,3286	0,6714	0,5217	0,4783	0,0638	0,9362	0,4991	0,1195	0,244218
temperature	<5	43	27	0,6143	0,3857	0,3023	0,6977	0,0741	0,9259	0,4218	0,1372	0,312046
temperature	<6	55	15	0,7857	0,2143	0,2364	0,7636	0,1333	0,8667	0,3610	0,2311	0,333160
temperature	<7	63	7	0,9000	0,1000	0,2222	0,7778	0,1429	0,8571	0,3457	0,2449	0,335601
temperature	<8	65	5	0,9286	0,0714	0,2308	0,7692	0,0000	1,0000	0,3550	0,0000	0,329670
temperature	<9	67	3	0,9571	0,0429	0,2239	0,7761	0,0000	1,0000	0,3475	0,0000	0,332623
temperature	<10	68	2	0,9714	0,0286	0,2206	0,7794	0,0000	1,0000	0,3439	0,0000	0,334034

Exceeded
15
Gini Impurity
0,336734694

$$Gini = 1 - \sum_j p_j^2 \qquad (2)$$

where pj is the probability of each of the decision-making subclasses in the candidate branch (exceeded, non-deferred).

Figure 1 shows a fragment of the decision tree for data.

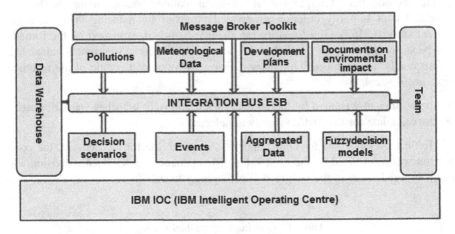

Fig. 1. Integral rail as integration technology in the IOC project (technologies selected on the basis of the RITMI model) (source: own study).

Analysis of Fig. 1 indicates a high temperature effect on PM10 dust. This dependence confirms the general theory about the influence of temperature on the dust concentration value. If, however, we assume that the presented example can be generalized, this analysis shows how important it is to equip the IoT node with the temperature sensor (its cost of delivery at the service of use) and to include this sensor in the construction of the business model. In this case, it is possible to design an IoT node with a group of sensors (pilot version), make a decision analysis of the impact of individual data obtained from these sensors on the measurement accuracy, estimate the cost of the IoT node and build a business model corresponding to the measurement accuracy assumed by the customer (Fig. 2).

The values of these data also indicate the construction and utility aspect of building a business model of using the IoT node. The design aspect indicates the need to place the temperature sensor outside the box containing other sensors and the microcontroller as well as connecting it with the microcontroller (the need to ensure high accuracy of temperature measurement). The utility aspect indicates the need to provide a local temperature measurement service.

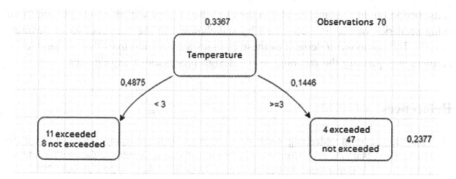

Fig. 2. The fragment of the decision tree for data

7 Summary

The article presents the method of constructing business models using decision trees for business processes analysis in IoT systems for Smart Cities. This method has been verified for the IoT system for PM10 dust measurement for the Gdańsk agglomeration. The presented example reflects, on the one hand, the influence of meteorological parameters on measurement accuracy, but on the other hand it supports decisions regarding the construction of the IoT measurement node.

In addition to the presentation of a detailed solution - the use of a decision tree for building a model and a business one for measuring PM10, the idea contained in the article has a general dimension. The solution proposed in the article can be used to construct any nodes/systems for Smart Cities using information systems for contact with residents. If we assume that the vision of Smart Cities requires the participation of residents in the decision-making processes, then building business models for the use of data obtained from such devices may become crucial for the development of the Smart Cities concept. In turn, in the construction of these models, the use of decision trees can significantly reduce the costs of designing IoT nodes and/or providing services resulting from the use of IoT nodes.

The authors of the work concentrated in the article only on the analysis of a simple case to show the context of contemporary city management. They presented both decision problems of these cities, applied information technology as well as devices. Therefore, the method verification had to be largely simplified.

Currently, the authors of the study are conducting research on the use of business models and their support in the processes of designing information systems. These studies take into account the place and role of open data for which the construction of business models seems more difficult. The construction of decision trees and their impact on the construction of business models is also important for the constructors of IoT devices. The authors think that the business process of building the IoT node is currently a viable business model. If the construction process is attached to the ESB bus, then in this way we will provide data that will correct the process of node construction on an ongoing basis. On the other hand, they will show the possibility of using scenarios of changes corrected by stakeholders in the construction of business models. The component/element of these changes are decision trees that are legible for partners and quantify the meaning in the design processes of any devices.

References

1. Manville, C., et al.: Mapping smart cities in the EU. In: Mapping Smart Cities in the EU/Catriona Manville et al. (Brussels: European Parliament, Directorate-General for Internal Policies, Policy Department A: Economic and Scientific Policy, p. 200, January 2014
2. Lee, J.H., Hancock, M.G., Hu, M.-C.: Towards an effective framework for building smart cities: lessons from Seoul and San Francisco 2014. In: Technological Forecasting and Social Change, vol. 89, pp. 80–99, November 2014
3. Business Model in Investopedia (2014). http://www.investopedia.com/terms/b/business model.asp
4. Sun, Y., Yan, H., Lu, C., Bie, R., Thomas, P.: A holistic approach to visualizing business models for the internet of things. Commun. Mobile Comput. 1, 1–7 (2012)
5. Bucherer, E., Eisert, U., Gassmann, O.: Towards systematic business model innovation: lessons from product innovation management. Creativity Innov. Manage. 21, 183–198 (2012)
6. Westerlund, M., Leminen, S., Rajahonka, M.: Designing business models for the internet of things. Technol. Innov. Manage. Rev. 4, 5–14 (2014)
7. Orłowski, C., Sarzyński, A., Karatzas, K., Katsifarakis, N., Nazarko, J.: Adaptation of an ANN-based air quality forecasting model to a new application area. In: Król, D., Nguyen, N. T., Shirai, K. (eds.) ACIIDS 2017. SCI, vol. 710, pp. 479–488. Springer, Cham (2017). https://doi.org/10.1007/978-3-319-56660-3_41
8. Orłowski, C., Sarzyński, A.: A model for forecasting pm 10 levels with the use of artificial neural networks. In: Information Systems Architecture and Technology—the use of IT Technologies to Support Organizational Management in Risky Environment, Wrocław (2014)
9. Borrego, C., et al.: Assessment of air quality microsensors versus reference methods: the EuNetAir joint exercise. Atmos. Environ. 147, 246–263 (2016). http://dx.doi.org/10.1016/j. atmosenv.2016.09.050
10. Jennings, P.: Managing the risks of smarter planet solutions. IBM J. Res. Develop. 54(4), 366–374 (2010)
11. Su, K., Li, J., Fu, H.: Smart City and the Applications. IEEE (2011)
12. Harrison, C., Donnelly, I.A.: A theory of smart cities. In: Proceedings of the 55th Annual Meeting of the ISSS (2011)

Author Index

Printed in the United States
By Bookmasters